SPACEBORNE WEATHER RADAR

For a complete list of the *Artech House Radar Library*, turn to the back of this book. . .

SPACEBORNE WEATHER RADAR

Robert Meneghini
and
Toshiaki Kozu

Artech House
Boston • London

Library of Congress Cataloging-in-Publication Data

Meneghini, R.
   Spaceborne weather radar / Robert Meneghini and Toshiaki Kozu.
     p. cm.
   Includes bibliographical references .
   ISBN 0-89006-382-6
   1. Radar meteorology. 2. Space-based radar--Design and
construction. 3. Precipitation (Meteorology)--Remote sensing.
I. Kozu, Toshi. II. Title.
QC973.5.M46 1990   89-49645
551.6'353--dc20   CIP

British Library Cataloguing in Publication Data

Meneghini, Robert
   Spaceborne weather radar.
   1. Weather. Forecasting. Use of spaceborne radar systems
   I. Title II. Kozu, Toshiaki
   621.6353

   ISBN 0-89006-382-6
   ISBN 13: 978-0-89006-382-8

International Standard Book Number: 0-89006-382-6
Library of Congress Catalog Card Number: 89-49645

10   9   8   7   6   5   4   3   2   1

Contents

Preface

This project began about two years ago with the intent not of writing a book but a chapter in the *Space-Based Radar Handbook* edited by Leopold Cantafio. Not only was the scope of our chapter too large, it did not fit the usual notion of a handbook contribution: it is, after all, difficult to write a practical guide to an instrument that has yet to be built. We were faced with the choice of expanding the chapter into a book or consigning it to oblivion. Evidently, we chose the former.

When we began writing about spaceborne weather radar there was, to our knowledge, only one proposal (the *Tropical Rain Measuring Mission* or TRMM) under active consideration. Since then there has been a proliferation of new proposals and studies. Like the TRMM radar, most of these have focused on the measurement of rain in the Tropics. However, the relatively modest objectives of the TRMM radar, limited to the sensing of rain with an incoherent, narrow swath radar, have expanded into studies of radars with one or more of the following features: wide swath, high resolution, dual-wavelengths, and doppler and cloud-sensing capabilities. Moreover, for missions that will focus on the radar monitoring of clouds as well as specific programs such as the *Global Energy and Water Cycle Experiment* (GEWEX), the exclusive interest in a low inclination orbiter has been broadened to include higher inclinations as well as polar orbiters. While these more ambitious studies will form the basis for the design of second generation spaceborne weather radar, there is also a growing interest in exploring simple, inexpensive monitoring of the rain and cloud by means of auxiliary modes on altimeters and synthetic aperture radars. Not all studies of spaceborne weather radars are recent. There was a recognition as early as the 1950s that radar is, in many respects, an ideal sensor for global cloud and rainfall monitoring. However, it is only in recent years that the technology has advanced enough to tap some of this potential. As important as the technological advances is the recognition of meteorologists, cloud physicists, and climatologists that the addition of the radar to microwave and VIS/IR radiometers represents both a natural evolution of space-based weather sensing and a necessary step in providing information required to advance the studies of global climate and tropical meteorology.

It is not a coincidence that the renewed interest in the measurement of clouds and precipitation from space comes as part of a much broader, international program to study global climate. From this perspective, precipitation is one of the parameters needed to understand the energy budget of the atmosphere, the hydrological cycle, and the large scale circulation of the atmosphere and oceans. The amount of the information required not only for these science objectives but also for immediate, direct applications to agriculture and water resource management is in striking contrast to our present ignorance of the distribution and amount of rainfall over much of the globe. As the name suggests, radar meteorology can be approached as either a special case of radar (Chapter 2) or as discipline within meteorology (Chapter 4). The complementary nature of the perspectives can also be understood by considering the remote sensing question from the point of view of the scattering medium or the sensor. If a discussion of radar meteorology were the only objective of the book, it would not only be superfluous, but also inadequate because many, more comprehensive books on radar meteorology, cloud physics and remote sensing already exist. The distinctive feature of this book is its concentration on those aspects of radar meteorology that are relevant to spaceborne measurements. This is particularly true of Chapter 3 and Chapter 5, which discuss design considerations and estimation methods as they apply to space-based radar sensing of precipitation.

The objective of the book is to provide sufficient information on spaceborne radar and meteorology so that the reader may understand the unique problems that spaceborne radar poses to the radar design engineer, the data analyst, and the meteorologist. Some of the specific issues that we have tried to address are:

- discussion of the radar design issues and the way in which the decisions affect radar performance;
- a description of the methods currently available for the estimation of precipitation characteristics from space;
- a review of the recent proposals for spaceborne weather radar and the relationship between measurement requirements and the radar design issues.

Some topics not discussed are spaceborne sensors that are complementary to the radar, such as lidar and VIS/IR and microwave radiometers. The beamfilling problem is touched on briefly, but is a more complex problem than we have indicated, especially in the case of broad-beam radars in heavy convective storms. Questions on the use of polarimetry and doppler from space are only now attracting the attention of radar meteorologists. The introductory material that we have presented is only a guide to work that will shortly supersede it.

We have tried to include at least a mention of the latest studies and proposals, perhaps to the extent that there is a danger of not distinguishing between the important and the ephemeral. However, we realize that we cannot avoid a certain amount of obsolescence in a field that is developing so rapidly. (We wish here to refer the interested reader to the series of workshop reports that will be issued by the World Meteorological Organization on the GEWEX Research Program in 1990). We believe that writing such a book is not a vain endeavor, however, if we are able to stimulate interest and to provide a framework for understanding further developments in this field.

The book relies heavily on work done by researchers at NASA/Goddard, the Communications Research Laboratory of Japan, the Jet Propulsion Laboratory, the Applied Physics Laboratory, and the Centre National d'Etudes Spatiales of France, as well as the work done at many universities and private companies. We wish to thank them for their cooperation. A note of thanks is also due to Dr. David Atlas, who kindly agreed to review the manuscript. We also wish especially to thank Jeffrey Jones for generating many of the illustrations that we have used.

Chapter 1
Introduction

Even before the launch of Sputnik in 1957, the satellite was considered an ideal platform for global observations of weather and climate. The scientific need for such data is as great as the applications are varied. Yet, more than three decades after the first studies of spaceborne weather radar appeared, no proposal has gone beyond the initial studies. Although much of the early work focused on radar, the radiometer (microwave, infrared and visible) has provided the only space-based observations to date of precipitation and clouds.

At least part of the reason for this lack of progress is due to the cost as well as the obstacles of weight, power, and reliability. Under such restrictions, there was some question as to whether an orbiting radar could supply data of an importance commensurate with its cost.

The relatively good prospect for a spaceborne weather radar to be launched during the 1990s is the result of several factors. Perhaps the most important development is the growing concern for the effect of human activity on the environment and on the need for a comprehensive program to monitor those processes which regulate the climate. As one of the basic components in the hydrological cycle, the global measurement of precipitation will advance the study of the coupling between ocean and atmosphere, aid in understanding the role of latent heating on large-scale atmospheric circulation, and improve our understanding of the morphology and dynamics of tropical storm systems. Alongside the science and modeling objectives are the immediate practical benefits to agriculture, water resource management, and flood warnings. To gauge the seasonal, annual, and long-term variations in the amount and distribution of rainfall, however, will require a steady, continuing commitment to developing, maintaining, and improving the capabilities of ground-based and spaceborne sensors.

The need for global precipitation data does not in itself warrant the use of spaceborne radar. To justify its use requires that it be technologically feasible and cost effective in the sense that its function cannot be duplicated by a more reliable or less expensive sensor. The unique features of radar are well recognized: unlike lidar, it can penetrate through rain and cloud; unlike the radiometer, it can vertically profile the rain and its sensitivity is not degraded by the high emissivities of a land background. While these benefits have long been recognized, in the absence of the necessary technology, they are merely potential advantages. Without the advances that have occurred in the efficiency and reliability of power amplifiers in the microwave and millimeter-wave region, in low-noise receivers and in antenna technology, spaceborne radar would suffer from poor sensitivity and reliability, and limited scanning capabilities.

One other reason for the renewed interest in spaceborne radar is a change in attitude which views the radar as one among a set of complementary sensors on the spacecraft. The sensors on the orbiter, in turn, are treated as one of many types of observation platform, which include airborne radars and radiometers, ground-based radars and rain gauge networks, and geosynchronous weather satellites. Although such a change in thinking does not solve any of the inherent limitations of spaceborne weather radar, it does provide a well defined role where its capabilities can be exploited even with a relatively modest implementation.

1.1 BRIEF SURVEY

Critical reviews of the early proposals for spaceborne weather radar are given in several technical and workshop reports [1–5]. Barrett and Martin [6] provide a thorough account of the history and issues up to the late 1970s. More recently, Hildebrand and Moore [7] devote a portion of the chapter on mobile platforms to recent designs and techniques for space-based weather radar.

The first studies on the feasibility of spaceborne weather radar occurred in the mid-1950s [8–10], followed by several papers on the question of the radar design [11–13]. This exploratory phase came to an end with the work of Dennis [1,2], who concluded that because of inherent as well as technological limitations, an orbiting weather radar would be of minimal value. After a decade of neglect in the US [14], a renewed interest and reassessment began in the mid-1970s [3–5, 15–18].

A conference, convened in 1980 on Precipitation Measurements from Space [19], helped to establish observational requirements for the applications of global precipitation data to climate and weather, agriculture, hydrology, and severe storms. Although more than five years elapsed between this conference and a number of recent proposals, the present work on spaceborne weather radar can be seen as a natural evolution of two ideas that emerged from earlier studies: that radar and radiometer combinations are needed to compensate for the measurement deficiencies of using either alone, and that a single low-earth-orbiting (LEO) platform is well suited to climatological applications where a statistical characterization of rainfall is useful even over long times and large areas.

Another important development is the growing interest of the international community in spaceborne weather radar. Radar designs for the Space Station Weather Radar as well as general system and space hardware requirements have been presented by Okamoto and colleagues [20–22]. Recently, a series of published papers provide the most detailed design of a spaceborne weather radar to date [23–28]. These papers, as well as a number of other studies in Japan and the US, have focused attention on radar designs for the Tropical Rain Measuring Mission and the Tropical Rain Mapping Radar for the Space Station [29–36]. A mission for studying the energy budget of the tropics has been proposed by CNET of France [37–38]. Other proposals and studies, which concentrated on cloud and light rain measurements, have included feasibility studies for the use of doppler radar [39–42]. Details of these and other radar designs can be found in Section 3.7 and throughout this book.

1.2 SOME ISSUES OF SPACEBORNE WEATHER RADAR

Despite the scientific benefits, and the innumerable applications of global precipitation, there was early recognition that most of the requirements far exceeded the technological capabilities. An outline of these problems can be stated in the following form [44].

(a) Because of antenna cost and size constraints, the use of nonattenuating wavelengths (>5 cm) would yield a field of view too broad to resolve important features of the precipitation. Moreover, the influence of the surface return, partial beam-filling and reflectivity gradients within the beam degrade its capabilities. As the severity of these problems increases at off-nadir angles, only narrow swaths are possible.

(b) Use of a shorter, attenuating wavelength (<3 cm) increases the resolution and, at the lighter rainfall rates, enhances the sensitivity. At moderate and high rainfall rates, however, the rapid increase in signal attenuation limits the depth of penetration into the storm.

(c) The demands of wide swath, good spatial resolution, and accuracy cannot all be satisfied within the practical limitations of satellite power.

(d) Over areas on the order of 100 km², the autocorrelation in the areal rainfall drops off rapidly within a span of several hours (Section 1.4). Thus, the crude temporal sampling from a single orbiting radar (once or twice daily) is insufficient to characterize rainfall amounts over such areas on a daily basis or to monitor the development of many storm systems.

The influence of these objections has been great. According to Matthews [3], "Dennis's findings essentially halted serious consideration of satellite-borne weather radars." Although the substance of many of these objections remains valid today not only for radar, but also for any rain or cloud sensor on an orbiting platform, responses to these criticisms have taken several forms: the assessment of the difficulties are pessimistic in light of new technologies; advances in the atmospheric sciences and the need for improved modeling and verification capabilities require global precipitation data; some of the limitations of the radar can be overcome by adding radiometric sensors and by combining both with the data provided by ground-based networks, airborne radars, and geosynchronous satellites.

Skolnik [4] noted that the seriousness of the surface clutter problem was somewhat overstated in the early studies, commenting that the decrease in the scattering cross section of the surface with incidence angle would allow the detection of fairly light rain rates. The rain detection problem therefore is not seriously degraded by the surface at off-nadir angles. However, to obtain quantitative estimates of rain rate, and even a coarse vertical profile, requires a narrow beamwidth in elevation.

Since 1963, significant progress has been made in radar, antenna, and spacecraft technology. For example, the weights and sizes of radar components have decreased, while transmitters are now available with higher efficiencies and greater reliability. The technologies for pulse compression and synthetic aperture radar have advanced to the point where such methods may be feasible for meteorological sensing. Adaptive scanning methods [45] that could be used to focus on portions of the storm of greatest interest are feasible only with large electronically scanned antennas.

Relative to most altimeters and scatterometers, the power requirements of a weather radar are large; moreover, because of the three-dimensional nature of the target and the desirability of profiling, the data rates will tend to be high and the processing requirements demanding. Although these requirements put severe demands on the spacecraft and computing capabilities of a decade ago, today they are considered well within the scope of present technology.

The importance of higher frequencies for spaceborne weather radars has prompted efforts to adapt attenuating-wavelength methods to orbiting platforms. Despite the inherent difficulties of doppler and polarimetry from space, the success of such methods for ground-based weather radars has led to studies for their use on a space platform. Data from the SIR-C [46] with its multiple polarization capability should be available within the next five years for an initial assessment of these methods. Radiometers on the weather satellites have shown the utility of proxy variables such as rain area and area-time integrals in the determination of area-average rain rate and total rainfall [6,19]. The application of these methods to spaceborne radar has stimulated renewed interest in this area [47–53].

In spite of these advances in technology and methodology, the prevailing philosophy is consistent with Dennis's advice to match the sensor to its capabilities. Indeed, the unique features of radar ensure its role in future spaceborne weather satellites.

1.3 INSTRUMENT COMPLEMENTARITY

The microwave radiometer has been shown to be useful for the quantitative determination of rain over the oceans, is more reliable and inexpensive than the weather radar, and generally can attain greater coverage. Although advances have been made in the use of microwave radiometers for rain estimation over land [54] and for some profiling capabilities with the use of multiple channels [55], the radar is best suited for profiling the storm and for quantitative determination of the rainfall rate over land. In fact, the radar measurements complement the radiometer by providing information on the maximum storm echo height, the presence and location of the melting layer, and in the case of dual-wavelengths or dual-polarization, the capability to distinguish between regions of rain and nonliquid hydrometeors. In cases where the radar and radiometer view a common volume of rain, profiles of the rain rate or liquid water content can be deduced by using the attenuation as measured from the radiometer to reconstruct the radar-derived profile of the rain rate [56].

Visible and infrared radiometers primarily measure not the precipitation but the thermal emissivity of the clouds. Rainfall amounts are generally deduced from either cloud indexing methods or lifetime histories (i.e., the change in cloud brightness temperature with time) together with information obtained from ground-based observations such as radar or rain gage networks. Although these methods are less direct than measurements in the microwave or millimeter-wave region, and are not useful for stratiform rains in the midlatitudes, the spatial and temporal resolution of the instruments is far greater than those attainable with microwave instruments. Because of this, they are presently the only spaceborne weather sensors possible from a geostationary orbit. These sensors and the more accurate but coarser resolution microwave instruments are complementary in that the latter can serve to calibrate the former by providing a correspondence between cloud brightness temperatures and the underlying precipitation field. This "initialization" can then be used to fill in those regions in space and time not covered by the LEO satellite. The complementarity of these three sensors is shown in Table 1.1 [35].

1.4 LARGE SCALE FEATURES OF PRECIPITATION

On average, about 50% of the earth's surface is covered by cloud and the ratio of the rain echo to the cloud area is seldom larger than 10% [57]. Although these estimates place an upper bound of 5% of the earth's surface receiving rain at a given time, the actual percentage is much less [6]. For the tropical zone between 30° N and 30° S, the occurrence of rain has been estimated to be about 4%, accounting for more than half the earth's rainfall [34]. Although contours of the

Table 1.1 TRMM Payload Complementarity [35]

	Microwave Radiometers	Radar	VIS/IR Radiometer
Advantages	Quantitative measure of rain	Quantitative measure of rain	Best spatial resolution
	Wide swath	Better spatial resolution	Distinguish between convective and stratiform precipitation
		Vertical profile of rain	Transfer standard to geosynchronous and to polar orbiters
		Can provide layer thickness	
		Works well over land	
Limitations	Less quantitative over land for low rainfall	Narrow swath	Less quantitative measure
	Moderate spatial resolution	Largely untested in space	Obscuration by cirrus shields

global mean annual precipitation have been constructed [6,35] over oceans and remote land areas, the annual rainfall at a particular location is probably known only to within a factor of 2 [34].

One of the most sought after characteristics of precipitation is the *rainfall rate* (or simply rain rate), R, defined as the volume of liquid water that falls through a unit area per unit time and normally expressed in units of mm per hour. Figure 1.1(a) [58] shows the globe divided into eight rain rate climate regimes. For each, the approximate distribution of rain rates is shown in Figure 1.1(b) *versus* the percentage of the time per year that the rain rate exceeds a specified value. For example, in the highest rainfall region (H) it is estimated that the rain rate exceeds 50 mm/hr 0.1% of the time or roughly 526 min per year, and exceeds a rain rate of 150 mm/hr for about 53 min annually.

Whereas most rain events are dominated by the lighter rainfall rates, the higher rain rates account for most of the total liquid water reaching the surface. The contribution from the various rain rate categories to the total rainfall can be found from the *probability density function* (pdf) for the rain rate, R. The relative contributions of the various rain rate categories to the total rainfall are obtained by multiplying the pdf by the rain rate, R, and integrating from zero to R. Figure 1.2 [52] shows the cumulative distributions of the total rainfall for three regions including the GATE (Global Atmospheric Research Program—Tropical Atlantic Experiment) area (a 280 km × 280 km area off the coast of Western Africa) for

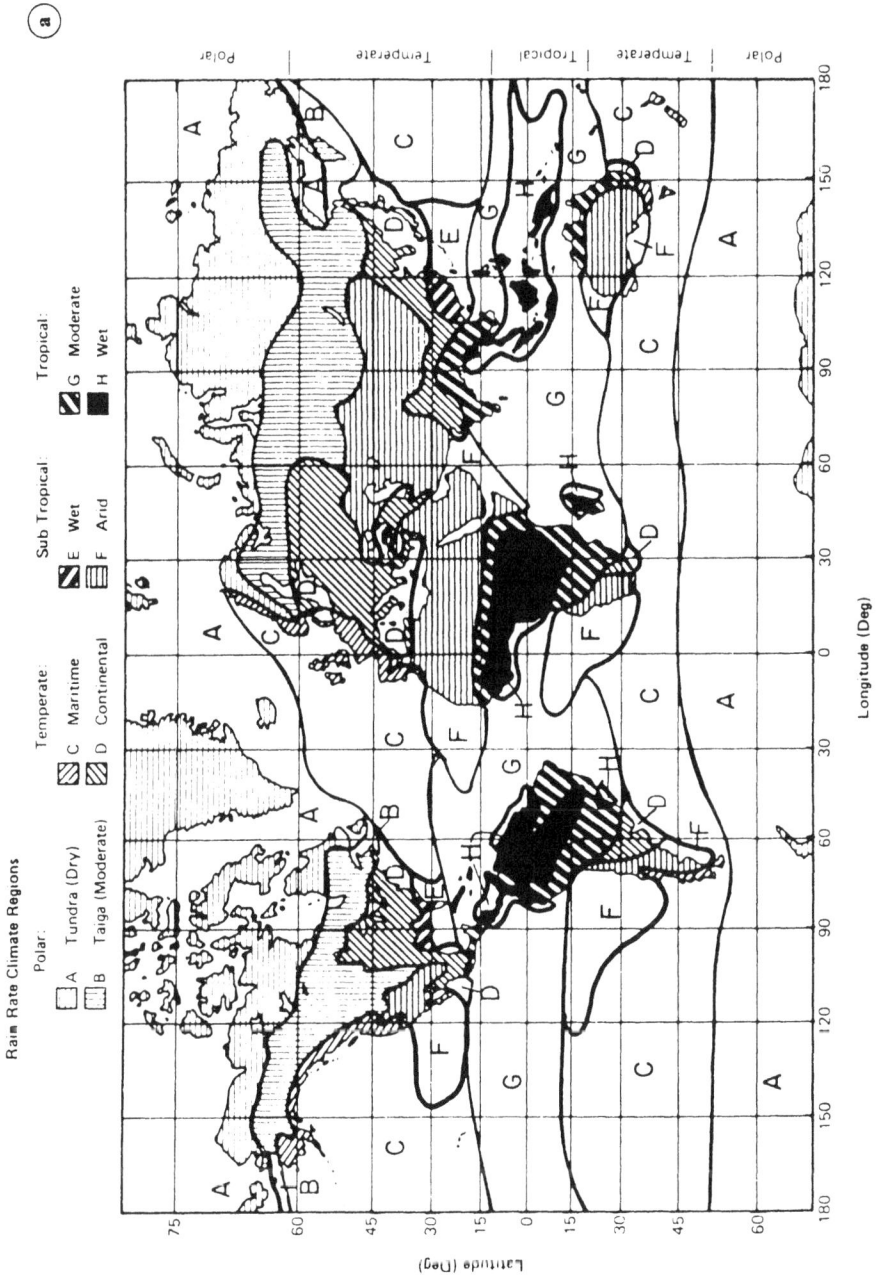

Figure 1.1 (a) Global rain rate climate regions; (b) Point rain rate distributions as a function of percent of year exceeded for the eight climate regions (from Crane and Blood [58]).

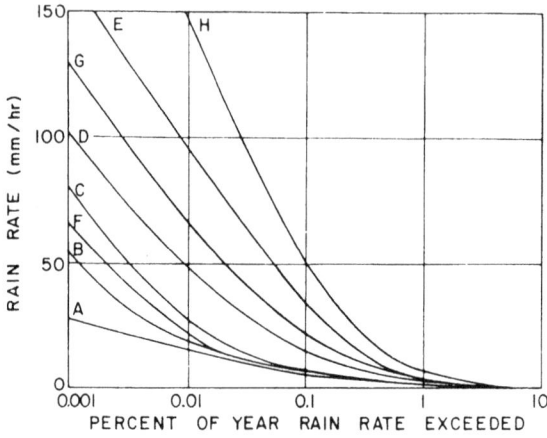

Fig. 1.1 continued

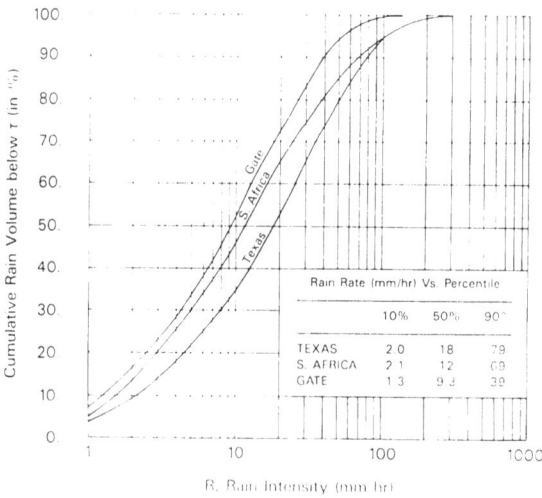

Figure 1.2 Percentage of the total rain as a function of the rain rate for GATE, Texas, and S. Africa for radar resolutions of approximately 1 km × 1 km × 15 min (from Rosenfeld *et al.* [52]).

approximately a 1 km × 1 km × 15 min sampling. The radar reflectivity factor (Z) and rainfall rate (R) relationships used to deduce R for the three regions are: $Z = 230 R^{1.25}$ (GATE); $Z = 383 R^{1.615}$ (Texas); $Z = 200 R^{1.4}$ (South Africa).

From the data shown in the figure, the dynamic range needed to characterize oceanic precipitation in the tropics (assuming that the GATE data sets are typical) is less stringent than over areas with intense local convection such as Texas. For example, a radar that can accurately estimate rain rates between 2 mm/hr and 40 mm/hr will account for about 84% of the total rainfall over the GATE area but only about 64% of the total rain volume over Texas. Fortunately, there appear to be excellent correlations between the fractional area over which the rain rate exceeds a certain threshold on one hand and the area-averaged rain rate on the other [51–53]. If relationships such as this can be tuned to the local climatology, then for those applications in which area-averaged rain rates are of primary importance, a large dynamic range may not be required (Section 5.6).

One of the most serious problems with the orbiting platform is the crude temporal sampling [59–61]. Using the GATE results, Laughlin [59] computed the normalized autocovariance function for the area-averaged rain rate for 10 different areas ranging from 4 km × 4 km to the full GATE area of 280 km × 280 km. Results for the 10 sampling areas are shown in Figure 1.3(a). Figure 1.3(b) shows the rms error as a function of the sampling interval for three values of the total observation time. These results indicate that the time rate of change of the area-averaged rain rate is inversely proportional to the size of the measurement area, so that as the spatial resolution is relaxed, the sampling frequency can be decreased [62–63]. Conversely, if the sampling interval and measurement area are fixed, the statistical characterization of the area-averaged rain rate can be improved by increasing the observation time. From the results of Figure 1.3, we can

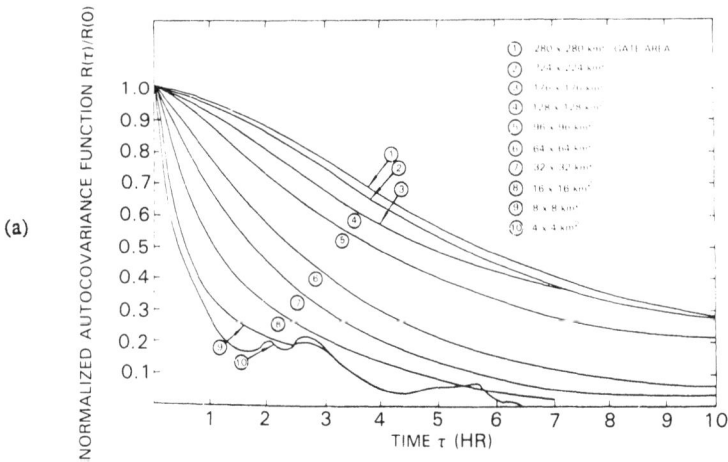

Figure 1.3 (a) Autocovariance functions of the mean area rain rate for 10 sampling areas within the Phase I GATE area; (b) normalized rms error (in percent) of the sample mean for a 280 km × 280 km area (from Laughlin [59]).

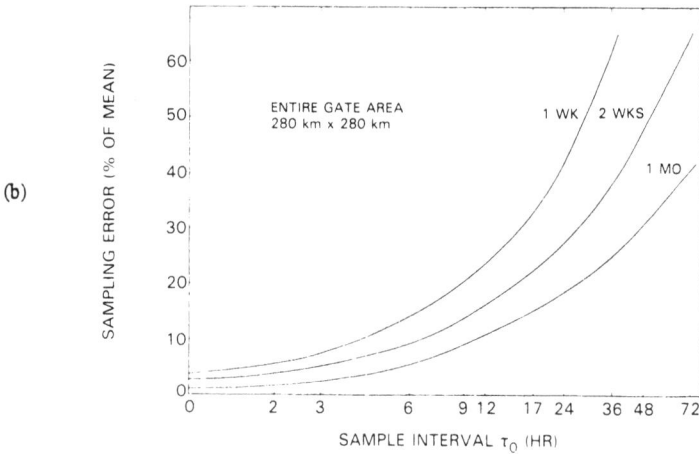

(b)

ENTIRE GATE AREA
280 km x 280 km

SAMPLING ERROR (% OF MEAN)

SAMPLE INTERVAL τ_0 (HR)

Fig. 1.3 continued

see that if the sensor observes a 280 km^2 area once per day, the error in the estimate of area-averaged rain rate deduced from samples taken in the course of a week is greater than 40% of the mean. If the sampling rate is increased to twice per day, the error in the sample mean is reduced to somewhat greater than 20% of the mean. For a total observation time of a month with two observations daily, the sampling error decreases to about 10%.

The issue of sampling can also be approached from the perspective of cell lifetimes [35,63]. Small scale cells (on the order of 5 km^2) have a typical lifetime of 15 min, whereas larger clusters of size 100 km^2 have average lifetimes of about 1 hr. The long temporal correlation of area-averaged rain rates over large areas suggests that the dissipation of a cell in one area is, on average, replaced by cell growth in another so that a steady state is achieved over times much longer than the typical lifetime of single cells or moderate cell clusters.

1.5 APPLICATIONS

The list of applications and measurement requirements compiled by the Precipitation Measurements Workshop [19] is given in Table 1.2. The horizontal resolutions that have been proposed for spaceborne weather radars generally fall between 2 and 6 km so that, with the possible exception of severe storms, the spatial requirements shown in Table 1.2 are quite feasible.

We must make a distinction between the resolution needed for a particular application and the *field of view* (FOV) of the instrument. Two-dimensional spectral analyses of precipitation fields yield a spatial correlation length on the order of

Table 1.2 Precipitation Data Requirements [19]

Application	Accuracy	Resolution Horizontal (km)	Temporal
1. Global climate			
Global	10–25%	200–500	1 week–1 month
Continent	10	25	1 day
2. Global weather	10	100	1 day
3. Synoptic weather forecast	10	100	6–12 hr
4. GCM	0.5–2 mm/day	100	1 day
5. Tropical cyclone (over water)	10–30%	2–20	0.5–6 hr
6. Thunderstorm-flash flood	10–30%	1–10	10–30 min
7. Mesoscale modeling	10–25%	25–100	15–60 min
8. Crop-yield modeling	10–30%	50	1 day
9. Soil-moisture evaluation	20%	20–100	1 day
10. Water-supply forecast	10%	10	1 day
11. Hydrological structure design	50%	10	1 week

6 km [63]. While this type of analysis gives a good indication of the spacing needed between adjacent fields of view, it does not in itself determine the instantaneous FOV. For example, a 6 km resolution produces a significant degradation in vertical resolution at large off-nadir angles; moreover, reflectivity gradients within the beam will degrade the accuracy of the rain rate estimate, especially because most radar methods are nonlinear in the sense that the measured quantity is not directly related to the spatial average of the precipitation parameter to be estimated. If a large antenna can be accommodated, one solution is the use of a narrow beam with a spacing between beams of 6 km or more: the narrow beam reduces the beam-filling and vertical smearing while the spacing between fields of view allows time for coverage and the gathering of a sufficient number of samples at each beam position.

While a radar on a low-earth orbiter generally can satisfy the spatial resolution requirements, with the important exception of global climate, the necessary temporal resolutions are impossible to achieve. Because of this limitation and because of the importance of large scale precipitation data, most of the recent proposals have focused on global climate. Table 1.3 shows the science and applications requirements of the TRMM [35]. Of the first five objectives, measurements are needed only over areas of 500 km² or larger with time scales ranging from biweekly to seasonally. For applications to tropical rain systems and dynamics (objectives 7 and 8), much finer temporal and spatial scales are needed. Although missions such as TRMM, TRAMAR, and BEST will not be continu-

Table 1.3 Precipitation Data Requirements for TRMM [35]

(1) *Climate Models and GCM Validation*
Space: $(500 \times 500$ km) or $\approx 10^5$ km^2
Time: Monthly mean
Accuracy: 1 mm/day (10 percent in heavy rain)
(2) *Documentation of Intraseasonal Variability (the 30- to 60-Day Oscillation—Signal ≈4 mm/day)*
Space: 10^6 km^2
Time: 15 days
Accuracy: 20 percent (0.8 mm/day)
(3) *Sahel Drought*
Space: 2.5×10^6 km^2 (2° latitude $\times$ 10° longitude)
Time: Seasonal
Accuracy: 20 percent
(4) *Monsoons*
Space: 10^6 km^2
Time: Monthly
Accuracy: 10 percent
(5) *Diurnal Cycle Over Ocean*
Space: 20° longitude, 5° latitude
Time: Bimonthly
Accuracy: 10 percent of first harmonic amplitude, 20 percent
                                of the second amplitude
(6) *Vertical Resolution (Portion of Swath Covered by Radar)*
For GCM's: 500 m (only possible out to about 6° scan angle, or
                                60 km swath)
For cloud statistics: 250 m (nadir only)
(7) *Tropical Rain Systems: Structure and Evolution*
Space: $\approx$20 km
Time: $\approx$12 hr
Accuracy: Much better than current ($\approx$30–50 percent)
Vertical resolution: 500 m to 4 km
(8) *Tropical Dynamics (Input to Regional Models and GCMs)*
Space: $\approx$20 km
Time: $\approx$12 hr
Accuracy: Much better than current ($\approx$30–50 percent)
Vertical resolution: 500 m to 4 km

ously able to monitor storm development, they will be able to study both large and small scale storms at different stages in their development. As noted earlier, the estimates of rain parameters at a particular area also can be used to calibrate observations from the geostationary weather satellites.

REFERENCES

[1]   Dennis, A.S., 1963: Rainfall Determinations by Meteorological Satellite Radar. NASA CR-50193, Stanford Research Institute, Menlo Park, CA.

[2]   Dennis, A.S., 1963: Fundamental Limitations on Precipitation Observations from Satellites. NASA CR-52848, Stanford Research Institute, Menlo Park, CA.

[3]   Matthews, R.E., ed., 1975: Active Microwave Workshop. NASA SP-376, 502 pp.

[4]   Skolnik, M.I., 1974: The Application of Satellite Radar for the Detection of Precipitation. NRL Report 2896, October, 100 pp.

[5]   Eckerman, J., and E.A. Wolf, 1975: Spaceborne meteorological radar measurement requirements meeting. NASA X-900-75-198, Goddard Space Flight Center, Greenbelt MD., 57 pp. plus appendices.

[6]   Barrett, E.C., and D.W. Martin, 1981: *The Use of Satellite Data in Rainfall Monitoring*, Academic Press, London, 340 pp.

[7]   Hildebrand, P.H., and R.K. Moore, 1989: Meteorological radar observations from mobile platforms in *Radar in Meteorology*, D. Atlas, ed., Amer. Meteor. Soc., Boston.

[8]   Wexler, H., 1954: Observing the weather from a satellite vehicle. *J. Brit. Interplanetary Soc.*, **13,** 269–276.

[9]   Wexler, H., 1957: The satellite and meteorology. *J. Astronaut.*, **4,** 1–6.

[10]   Widger, W.K., Jr., and Touart, C.N., 1957: Utilization of satellite observations in weather analysis and forecasting. *Bull. Amer. Meteor. Soc.*, **38,** 521–533.

[11]   Mook, C.P., and D.S. Johnson, 1959: A proposed weather radar and beacon system for use with meteorological earth satellites. *Proc. 3rd Nat. Convention on Military Electronics*, Washington, DC, 29 June–1 July, 206–209.

[12]   Keigler, J.E., and L. Krawitz, 1960: Weather radar observations from an earth satellite. *J. Geophys. Res.*, **65,** 2793–2808.

[13]   Katzenstein, H., and H. Sullivan, 1960: A new principle for satellite-borne meteorological radar. *Proc. 8th Weather Radar Conf.*, Amer. Meteor. Soc., Boston, 505–515.

[14]   Stepanenko, V.D., 1966: Radar in Meteorology. Gidrometeoizdat (Leningrad). (Partial translation available from Joint Publications Research Service, Washington, DC.)

[15]   Katz, I., 1975: Active microwave sensing of the atmosphere from satellites. *Proc. 16th Conf. on Radar Meteorology*, Amer. Meteor. Soc., Boston, 246–252.

[16]   Eckerman, J., 1975: Meteorological radar facility for the space shuttle. *IEEE National Telecomm. Conf.*, New Orleans, IEEE Publ. 75 CH1015 CSCB, 37-6–37-17.

[17]   Eckerman, J., R. Meneghini, and D. Atlas, 1978: Average Rainfall Determination from a Scanning Beam Spaceborne Radar. NASA Tech. Memo. 79664, 35 pp plus appendices.

[18]   Goggins, W.B. Jr., 1974: Satellite Weather Radar. AFCRL-TR-74-0479.

[19]   Atlas, D., and O.W. Thiele, eds., 1981: Precipitation measurements from space. Workshop Report, NASA/Goddard Space Flight Center, Greenbelt, MD.

[20]   Okamoto, K., S. Miyazaki, and T. Ishida, 1979: Remote sensing of precipitation by a satellite-borne microwave remote sensor. *Acta Astronautica*, **6,** 1043–1060.

[21]   Okamoto, K., T. Ojima, S. Yoshikado, H. Masuko, H. Inomata, and N. Fugono, 1982: System design and examination of spaceborne microwave rain-scatterometer. *Acta Astronautica*, **9,** 713–721.

[22]   Okamoto, K., 1983: Remote sensing of precipitation by weather radar system at space station. Preprints *21st Conf. on Radar Meteor.*, Amer. Meteor. Soc., Boston, 263–269.

[23]   Okamoto, K., 1988: A feasibility study of the rain radar for the tropical rainfall measuring mission: 1. Introduction. *J. Comm. Research Lab.*, **35,** 109–110.

[24]   Awaka, J., T. Kozu, and K. Okamoto, 1988: A feasibility study of the rain radar for the tropical rainfall measuring mission: 2. Determination of basic system parameters., *J. Comm. Research Lab.*, **35,** 111–133.

[25]   Nakamura, K., and T. Ihara, 1988: A feasibility study of the rain radar for the tropical rainfall measuring mission: 3. Radar type and antenna. *J. Comm. Research Lab.,* **35,** 135–148.

[26]   Ihara, T., and K. Nakamura, 1988: A feasibility study of the rain radar for the tropical rainfall measuring mission: 4. A discussion of pulse-compression and adaptive scanning. *J. Comm. Research Lab.,* **35,** 149–161.

[27]   Manabe, T., and T. Ihara, 1988: A feasibility study of the rain radar for the tropical rainfall measuring mission: 5. Effects of surface clutter on rain measurements from satellite. *J. Comm. Research Lab.,* **35,** 163–181.

[28]   Okamoto, K., J. Awaka, and T. Kozu, 1988: A feasibility study of the rain radar for the tropical rainfall measuring mission: 6. A case study of rain radar systems. *J. Comm. Research Lab.,* **35,** 183–208.

[29]   Okamoto, K., T. Kozu, K. Nakamura, and T. Ihara, 1988: Tropical rainfall measuring mission radar. *Tropical Rainfall Measurements,* J.S. Theon and N. Fugono, eds. A Deepak Publishing, Hampton, VA., 213–219.

[30]   Im, K.E., F.K. Li, W.J. Wilson, and D. Rosing, 1987: Conceptual design of a spaceborne radar for global rain mapping. *Proc. IGARSS,* Ann Arbor, MI.

[31]   Li, F., K.I. Im, W.J. Wilson, and C. Elachi, 1988: On the design issues for a spaceborne rain mapping radar. *Tropical Rainfall Measurements,* J.S. Theon and N. Fugono, eds. A Deepak Publishing, Hampton, VA, 387–393.

[32]   Im, E., and F. Li, 1989: Tropical rain mapping radar on the Space Station. *IGARSS '89,* Vancouver, British Columbia, July 10–14, 1485–1490.

[33]   Goldhirsh, J., 1988: Analysis of algorithms for the retrieval of rain rate profiles from a spaceborne dual-wavelength radar. *IEEE Trans. Geosci. and Remote Sens.,* **GE-26,** 98–114.

[34]   Thiele, O.W., ed., 1987: On requirements for a satellite mission to measure tropical rainfall. NASA Ref. Publ. #1183, 49 pp.

[35]   Simpson, J., ed., 1988: TRMM—A satellite mission to measure tropical rainfall. Report of the Science Steering Group. NASA/GSFC, August, 94 pp.

[36]   Simpson, J., R.F. Adler, and G.R. North, 1988: A proposed tropical rainfall measuring mission. *Bull. Amer. Meteor. Soc.,* **69,** 278–295.

[37]   BEST: Tropical System Energy Budget, 1988. *Cenᵗre National d'Etudes Spatiales,* October, 58 pp.

[38]   Marzoug, M., P. Amayenc, J. Testud, and N. Karouche, 1989: Conceptual design of the spaceborne rain radar of the B.E.S.T. project. Preprints *24th Conf. on Radar Meteor.,* March 27–31, Amer. Meteor. Soc., Boston, 597–600.

[39]   Lhermitte, R., 1981: Satellite borne dual millimetric wave length radar. In Precipitation Measurements from Space: Workshop Report, Atlas, D., and O.W. Thiele, eds., NASA/GSFC, Greenbelt, MD, pp. D-277–D-282.

[40]   Lhermitte, R., 1989: Satellite-borne millimeter wave Doppler radar. *URSI Commission F, Open Symposium,* La Londe-Les-Maures, France, September 11–15.

[41]   Nathanson, F.E., T.H. Slocumb, L. Brooks, R.K. Crane, and S.W. McCandless, 1988: Radar Sounder. Final Report, Contract Number F19628-87-C-0231, 133 pp.

[42]   Nathanson, F.E., T.H. Slocumb, S.W. McCandless, and R.K. Crane, 1989: A space based radar to measure clouds and rain. *IGARSS '89,* Vancouver, Canada, July 10–14, 1484.

[43]   Im, K.E., and D. Atlas, 1988: The estimation of precipitation intensity in the presence' of surface backscatter and the converse using a spaceborne radar. *Tropical Rainfall Measurements,* J.S. Theon and N. Fugono, eds., A Deepak Publishing, Hampton, VA., 221–227.

[44]   Jorgensen *et al.,* 1989: Panel Report for the Airborne/Spaceborne Radar Panel Technology Session in Radar in Meteorology, D. Atlas, ed., Amer. Meteor. Soc., Boston.

[45]   Atlas, D., 1982: Adaptively pointing spaceborne radar for precipitation measurements. *J. Appl. Meteor.,* **21,** 429–431.

[46]   NASA-JPL, 1986: Shuttle Imaging Radar-C Science Plan. JPL Publication 86–29, September.

[47]   Crane, R.K., 1981: Sampling problems: The small scale structure of precipitation. In Precipitation Measurements from Space: Workshop Rept., Atlas, D., and O.W. Thiele, eds., NASA/GSFC, Greenbelt, MD, pp. D-41–D-49.

[48]   Crane, R.K. and K.R. Hardy, 1981: The HIPLEX program in Colby-Goodland Kansas: 1976–1980, Rep. P-1552-f, 144 pp., Environmental Research and Technology, Inc., Concord, MA.

[49]   Doneaud, A.A., P.L. Smith, A.S. Dennis, and S. Sengupta, 1981: A simple method for estimating convective rain volume over an area. Water Resources Research, 17, 1676–1682.

[50]   Doneaud, A.A., S. Ionescu-Niscov, D.L. Priegnitz, and P.L. Smith, 1984: The area-time integral as an indicator for convective rain volumes. J. Clim. Appl. Meteor., 23, 555–561.

[51]   Chiu, L.S., 1988: Rain estimation from satellites: Area rainfall-rain area relation. Third Conf. Satellite Meteor. and Oceanog., Feb. 1–5, Amer. Meteor. Soc., Anaheim, CA, 363–368.

[52]   Rosenfeld, D., D. Atlas, and D.A. Short, 1988: The estimation of convective rainfall by area integrals, part II: The height area rainfall threshold (HART) method. Conf. on Mesoscale Precip., Cambridge, MA, September 13–17.

[53]   Atlas, D., D. Rosenfeld, and D.A. Short, 1988: The estimation of convective rainfall by area integrals. Part I: The theoretical and empirical basis. Conf. on Mesoscale Precipitation, MIT, Cambridge, MA., September 13–17.

[54]   Spencer, R.W., D.W. Martin, B.B. Hinton, and J.A. Weinman, 1983: Satellite microwave radiances correlated with radar rain rates over land. Nature, 304, 141–143.

[55]   Weinman, J.A., C.D. Kummerow, and C.S. Atwater, 1988: An algorithm to derive precipitation profiles from a downward viewing radar and a multifrequency passive radiometer. IGARSS, September, 13–16.

[56]   Weinman, J.A., R. Meneghini, and K. Nakamura, 1989: Comparison of rainfall profiles retrieved from dual-frequency radar and from combined radar and passive microwave radiometric measurements. Fourth Conf. on Satellite Meteor. and Oceanography. May 16–19, San Diego, CA., 27–30.

[57]   Griffith, C.G., W.L. Woodley, P.G. Grube, D.W. Martin, J. Stout, and D.N. Sikdar, 1978: Rain estimation from geosynchronous satellite imagery—visible and infrared studies. Mon. Wea. Rev., 106, 1153–1171.

[58]   Crane, R.K., and D.W. Blood, 1979: Handbook for the estimation of microwave propagation effects—link calculations for earth-space paths. Rep. P-7376-TR1, 80 pp., Environmental Research and Technology, Inc., Concord, MA.

[59]   Laughlin, C.R., 1981: On the effect of temporal sampling on the observation of mean rainfall. In Precipitation Measurements from Space: Workshop Report, Atlas, D., and O.W. Thiele, eds., NASA/GSFC, Greenbelt, MD, pp. D-59–D-66.

[60]   North, G.R., and S. Nakamoto, 1989: Comparing rain estimation designs. Fourth Conf. on Satellite Meteor. and Oceanography, May 16–19, San Diego, Amer. Meteor. Soc., Boston, 24–26.

[61]   Bell, T.L., 1987: A space-time stochastic model of rainfall for satellite remote-sensing studies. J. Geophys. Res. Atmos., 92, 9631–9634.

[62]   Flueck, J.A., 1981: Some statistical problems inherent in measuring precipitation. In Precipitation Measurements from Space: Workshop Rept., Atlas, D., and O.W. Thiele, eds., NASA/GSFC, Greenbelt, MD, pp. D-50–D-58.

[63]   Crane, R.K., 1981: Sampling problems: The small scale structure of precipitation. In Precipitation Measurements from Space: Workshop Rept., Atlas, D., and O.W. Thiele, eds., NASA/GSFC, Greenbelt, MD, pp. D-41–D-49.

Chapter 2
Radar Equations

In assessing the performance of a spaceborne weather radar, the first concern is the detectability of the target over the range of expected conditions. This depends not only on the signal strength from the hydrometeors, but also on the magnitude of the interference sources such as surface return and system noise. Moreover, because the relative positions of the hydrometeors are constantly changing, the return echo must be averaged to obtain a reliable estimate of the mean radar reflectivity within the resolution volume. In the following sections, we discuss several forms of the meteorological radar equation and the interpretation of the parameters for various radar designs. This is followed by expressions for the surface return, the "mirror-image" return, and the system noise. Following a discussion of the statistical properties of the signals, modifications to the radar equation for bistatic geometry and for polarimetry methods are presented. The chapter concludes with a discussion of doppler considerations for spaceborne weather radars.

2.1 METEOROLOGICAL RADAR EQUATIONS

The equation that relates the radar received power to the characteristics of the scattering medium has a number of different forms. For the sensing of precipitation, the fractional volume occupied by the particles is generally much less than one; moreover, the absorption cross section is generally much larger than the scattering cross section. Assuming that the particles are distributed randomly within the pulse volume, the scattered powers from individual particles can be summed algebraically (incoherently). To specify the equation in more detail, we often assume that the resolution volume is sufficiently small so that the properties of the rain or cloud are constant over the integration volume.

A heuristic derivation of the radar equation begins by noting that the power density, S_i, incident on an incremental volume ΔV in the far-field of the antenna is (see Figure 2.1):

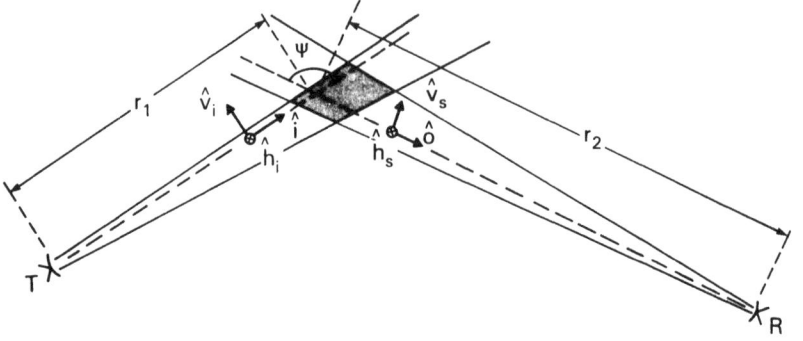

Figure 2.1 Bistatic scattering geometry from the transmitter, T, to the receiver. R.

$$S_i(t) = (P_0/4\pi r_1^2)|u(t - r_1/c)|^2 G_t(\hat{\imath})e^{-\gamma(r_1)} \tag{2.1}$$

where

P_0 = peak transmitted power,
$u(t)$ = complex voltage envelope of the transmitted pulse,
$G_t(\hat{\imath})$ = antenna gain pattern on transmit along direction $\hat{\imath}$,
γ = optical depth,
r_1 = range from antenna to ΔV,
c = speed of light in free space.

The antenna gain pattern can be expressed as

$$G_t = G_0 g^2(\hat{\imath}) \tag{2.2}$$

where G_0 is the antenna gain and $g(\hat{\imath})$ is the ratio of the electric field in the direction $\hat{\imath}$ to the maximum field. In either form, the gain pattern represents the angular distribution of power, while the range dependence is given by the product of the free-space loss $(4\pi r_1^2)^{-1}$, the envelope of the transmitted waveform $P_0|u(t - r_1/c)|^2$, and the attenuation out to the range r_1, $\exp(-\gamma(r_1))$. The optical depth γ represents the loss in power in the direction of propagation caused by absorption and scattering. In meteorological applications, it is normally expressed in terms of a specific attenuation k (dB per unit length), where

$$\gamma(r) = 0.1 \ln 10 \int_0^r k(s) \, ds \tag{2.3}$$

and

$$k(s) = 4.343(1/\Delta V') \sum_j \sigma_{tj} \tag{2.4}$$

where σ_{tj} is the extinction or total cross section of the jth particle and where the summation extends over all particles within an incremental volume $\Delta V'$ along the path centered at range s.

The power density S_s scattered from ΔV into the direction $\hat{o}$ at a distance r_2 from ΔV can be written as (see Figure 2.1):

$$S_s(t) = S_i(t - r_2/c)[(1/\Delta V) \sum_j \sigma_{bj}(\hat{o},\hat{\imath})](\Delta V/4\pi r_2^2)e^{-\gamma(r_2)} \tag{2.5}$$

where $\sigma_{bj}(\hat{o},\hat{\imath})$ is the bistatic radar cross section of the jth particle along the direction $\hat{o}$ for an incident wave along $\hat{\imath}$. To represent the scattering properties of the assemblage of particles in ΔV, a bistatic radar reflectivity is introduced:

$$\eta(\hat{o},\hat{\imath}) = (1/\Delta V) \sum_j \sigma_{bj}(\hat{o},\hat{\imath}) \tag{2.6}$$

As with the specific attenuation, the dimensions of η are area (cross section) per unit volume.

The power available at the receiving antenna, $P_r(t)$, is a product of S_s and the effective antenna area A_e:

$$P_r(t) = S_s(t)A_e \tag{2.7}$$

When the antenna is matched in impedance to the load and the polarization is matched to the incoming field, the relationship between A_e and the gain of the receiving antenna, $G_r(\hat{o})$ is [1]:

$$A_e = (\lambda^2/4\pi)G_r(\hat{o}) \tag{2.8}$$

where λ is the wavelength. This equation can be used even when losses or mismatches are present if $G_r(\hat{o})$ is defined as 4π times the power per steradian in the direction $\hat{o}$ over the maximum available power from the generator (using the receiving antenna as a transmitter). This differs from the antenna directivity by the ratio of the power that is radiated to the available power at the generator.

Combining the preceding relations and integrating over all space gives

$$P_r(t) = \frac{\lambda^2 P_0}{(4\pi)^3} \int_V \frac{G_t(\hat{\imath})G_r(\hat{o})\eta(\hat{o},\hat{\imath})e^{-\gamma(r_1)-\gamma(r_2)}}{r_1^2 r_2^2} |u(t - (r_1 + r_2)/c)|^2 \, dV \tag{2.9}$$

A more rigorous derivation of (2.9) shows that it follows as a consequence of a narrow-band radar, where the scattering medium can be characterized as time-invariant and linear with uncorrelated scattering [2].

Monostatic Radar

By far the most common situation occurs when the same antenna is used for transmitting and receiving. In this case, $G_t(\hat{\imath}) = G_r(-\hat{\imath}) = G$, $r_1 = r_2 = r$ and $\eta(-\hat{\imath},\hat{\imath}) = \eta$, where η is defined as the radar reflectivity. Letting $dV = r^2\, dr\, d\Omega$, (2.9) becomes

$$P_r(t) = \frac{\lambda^2 P_0}{(4\pi)^3} \int_V \frac{G^2(\Omega)\eta(r,\Omega)e^{-2\gamma(r,\Omega)}}{r^2}\, |u(t - 2r/c)|^2\, d\Omega\, dr \tag{2.10}$$

If most of the energy is concentrated into a beam narrow enough that the radar reflectivity and attenuation are approximately independent of the angular variations, the equation can be factored into a product of integrals over range and solid angle:

$$P_r(t) = \frac{\lambda^2 P_0}{(4\pi)^3} \int_\Omega G^2(\Omega)\, d\Omega \int_r \frac{\eta(r)e^{-2\gamma(r)}}{r^2}\, |u(t - 2r/c)|^2\, dr \tag{2.11}$$

To evaluate the angular integral, a standard approximation is to assume that the distribution of power within the main beam can be described by a two-dimensional Gaussian, and that the sidelobes can be neglected. For this "pencil-beam approximation" [3]:

$$G(\Omega) = G_0 \exp\{- \ln2\, [(2\theta/\theta_B)^2 + (2\phi/\phi_B)^2]\} \tag{2.12}$$

where θ_B, ϕ_B are the half-power beamwidths along the two principal directions of the elliptically shaped main beam. The azimuthal angles θ, ϕ are measured from the point of maximum gain. In the case of a circular main beam, $G(\Omega)$ can be written as $G_0 \exp[- \ln2\, (2\theta/\theta_B)^2]$, where θ is the polar angle measured from the point of maximum gain.

A somewhat different Gaussian approximation has been considered by Stephens [4]:

$$G(\Omega) = G_0 \exp\{-4 \ln2\, \theta^2[(\cos\, \phi/\Phi_0)^2 + (\sin\, \phi/\Phi_1)^2]\} \tag{2.13}$$

where θ is the polar angle measured from the direction of maximum gain and ϕ is the azimuthal angle. The half-power beamwidths Φ_0, Φ_1 are aligned along the $\phi = 0$ and $\phi = \pi/2$ planes, respectively. Integration of G^2 over all angles gives a result that agrees to within 0.1 dB of the standard result as long as the ratio of beamwidths is less than 5. For beamwidth ratios of 7 and 8.88, the integration of G^2 is smaller by 0.5 dB and 1 dB, respectively, than the standard result.

The gain G_0 is normally a known parameter of the radar. For purposes of relating the return power to the radar resolution, G_0 is often expressed as a function of the beamwidths. Following [1,3], G_0 is related to the optimum gain G_{0p} and the physical antenna area A_p by

$$G_0 = \alpha\beta G_{0p} = 4\pi\alpha\beta A_p/\lambda^2 \qquad (2.14)$$

where α is the fraction of power transmitted by the feed that ends up as radiated power and β is related to the spatial distribution of the primary field on the antenna aperture or reflector surface. The beamwidth product can be expressed in terms of A_p and a function of the efficiency factor β, for example $h^2(\beta)$, as

$$\theta_B\phi_B = h^2(\beta)\pi\lambda^2/4A_p \qquad (2.15)$$

Combining these relations gives

$$G_0 = (\pi p)^2/\theta_B\phi_B \qquad (2.16)$$

where $p^2 = \alpha\beta h^2(\beta)$. Probert-Jones [3] has compared this to a well known class of gain patterns given by

$$G_q(\theta) = \frac{\pi^2\alpha\beta D_q^2}{\lambda^2}\left[\frac{2^q(q+1)!J_{q+1}(u)}{u^{q+1}}\right]^2 \qquad (2.17)$$

where D_q is the antenna diameter, J_p is the Bessel function of order p, $u = \pi D_q/\lambda \sin\theta$ and where θ is the polar angle measured from the maximum of the pattern. The gain patterns correspond to a field distribution over a circular aperture of the form $[(D_q/2)^2 - \rho^2]^q$ where ρ is the radial distance measured from the center of the aperture and q is a nonnegative integer. To compare $G_q(\theta)$ as a function of q, the diameter is adjusted to give the same 3 dB beamwidth for all q. Letting I_q be the integral of $G_q^2(\theta)$ over all θ, I_q is found to vary by less than 0.64 dB over the full range of q. The factor $p^2 = \alpha\beta h^2(\beta)$, moreover, ranges from a minimum of 1.04α at $q = 0$ to a maximum of about 1.21α at $q = 2$, slowly decreasing thereafter to 1.124α for large q. Because α is typically between 0.8 and 0.9 for a parabolic antenna, then $p \doteq 1$ and $G_0 = \pi^2/\theta_B^2$ or more generally, $G_0 = \pi^2/\theta_B\phi_B$.

Substituting (2.12) into (2.11) and integrating gives

$$P_r(t) = \frac{P_0(G_0\lambda)^2\theta_B\phi_B}{2^9\pi^2\ln2}\int_r \frac{e^{-2\gamma(r)}\eta(r)}{r^2}|u(t - 2r/c)|^2\,dr \qquad (2.18)$$

In evaluating the range integration, the waveform shape is taken to be rectangular with duration τ. If the spatial extent of the transmitted pulse is small with respect to variations in η and r is approximately constant over the pulse length, then,

$$P_r(r) = \frac{P_0(G_0\lambda)^2\theta_B\phi_B}{2^{10}\pi^2 \ln 2\ r^2}\ \eta(r)c\tau e^{-0.2\ \ln 10\ \int_0^r k(s)\ ds} \tag{2.19}$$

where r, the midpoint of the pulse volume, is related to t by $r = (c/2)(t - \tau/2)$, and where the optical depth γ has been written in terms of the specific attenuation, k, using (2.3).

For Rayleigh scattering of spherical water droplets, η is given by

$$\eta = (\pi^5/\lambda^4)|K_w|^2(1/\Delta V) \sum_i D_i^6 \tag{2.20}$$

where D_i is the drop diameter of the ith particle within the volume ΔV, and where the dielectric factor K is given by

$$K = (m^2 - 1)/(m^2 + 2) \tag{2.21}$$

where m is the complex index of refraction. The notation K_w indicates that the refractive index of water is to be used. The radar reflectivity factor, Z, is defined in terms of the Rayleigh reflectivity of water drops by [5]:

$$Z = (\lambda^4\eta/\pi^5|K_w|^2) = (1/\Delta V) \sum_i D_i^6 \tag{2.22}$$

where the latter equality follows from (2.20). For a continuous distribution of drop sizes with $N(D)$ as the number density of drops per unit diameter, Z can be written as

$$Z = \int_D D^6 N(D)\ dD \tag{2.23}$$

This definition can be extended to non-Rayleigh scattering by replacing Z with an effective or equivalent radar reflectivity Z_e where Z_e follows from the general definition of η, (2.6), with $\hat{o} = -\hat{i}$. For a continuous distribution of drop sizes, Z_e is

$$Z_e = \frac{\lambda^4}{\pi^5|K_w|^2} \int_D \sigma_b(D)N(D)\ dD \tag{2.24}$$

where $\sigma_b(D)$ is the radar cross section. For Rayleigh scattering, $Z_e = Z$ for water drops and $Z_e = (|K|^2/|K_w|^2)Z$ for nonliquid hydrometeors. The return power can now be expressed as a product of a radar "constant" and terms that depend only on the characteristics of the scattering medium:

$$P_r(r) = \frac{C|K_w|^2 Z_e}{r^2} e^{-0.2 \ln 10 \int_0^r k(s)\, ds} \tag{2.25}$$

where the radar constant is

$$C = (\pi^3/2^{10} \ln 2) P_0 G_0^2 \theta_B \phi_B c\tau/\lambda^2 \tag{2.26}$$

To include losses on transmission, l_t, and reception, l_r, (l_t, $l_r < 1$) the expression for C is to be multiplied by $l_t l_r$. In many cases, it is useful to write the return power in terms of meteorological quantities such as the rainfall rate, R. If we express Z_e and k as a function of R in the usual power law forms given by $Z = aR^b$ and $k = \alpha R^\beta$, where the units of Z_e, k and R are given by mm^6/m^3, dB/km, and mm/h, respectively, then the radar equation becomes

$$P_r(r) = \frac{4.37 \times 10^{-20} P_0 G_0^2 \theta_B \phi_B c\tau |K_w|^2 aR^b}{r^2 \lambda^2} e^{-0.2 \ln 10 \int_0^r \alpha R^\beta\, ds} \tag{2.27}$$

where P_r, P_0 are in watts, and $c\tau$, and λ are in meters. The range term in the coefficient of (2.27) is in meters and that appearing in the exponent is in km.

In the following sections, we discuss some of the modifications of the basic meterological radar equation.

Beamfilling Factor

In cases where only a fraction of the pulse volume is occupied by the scatterers, a correction factor can be introduced. *Fractional beamfilling* is a special case of the general problem in which the reflectivity and attenuation vary as a function of position within the pulse volume. For a spaceborne weather radar, the problem of partial beamfilling is further compounded by the surface and the storm top, especially for broad-beam antennas and incidence angles directed away from nadir.

For the partial beamfilling geometries shown in Figure 2.2, the multiplicative correction factor to the radar equation can be approximated by

$$F = \int_P G^2\, dS \Big/ \int_F G^2\, dS \tag{2.28}$$

where the integral in the numerator (partial filling) extends only over the raining (shaded) portion of the pulse volume, while the integration in the denominator is over the full pulse volume. In this approximation, the variation in range is neglected; moreover, the reflectivity and attenuation are assumed to be constant

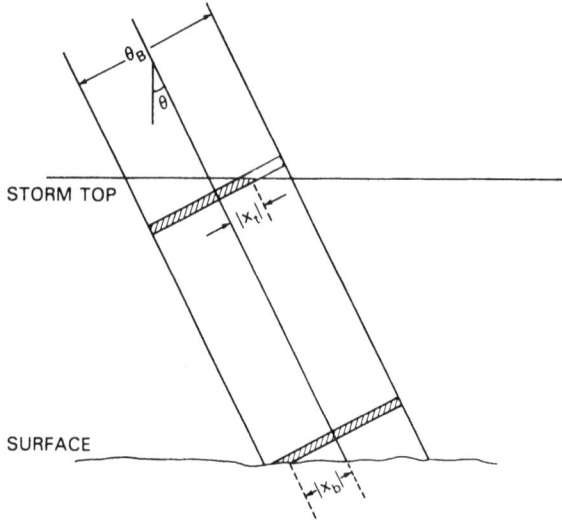

Figure 2.2 Schematic of partial beamfilling at the storm top and surface.

over the partially filled volume. Using a rotationally symmetric Gaussian gain pattern, then,

$$F = 0.5[\text{sgn}(x_t)\Phi(q|x_t|) - \text{sgn}(x_b)\Phi(q|x_b|)] \tag{2.29}$$

with

$$x_t = r\{\cos^{-1}[(H - H_s)/r] - \theta\} \tag{2.30}$$

$$x_b = r[\cos^{-1}(H/r) - \theta] \tag{2.31}$$

$$q = \sqrt{8 \ln 2}/r\theta_B \tag{2.32}$$

$$\Phi(u) = (2/\sqrt{\pi}) \int_0^u \exp(-t^2)\, dt \tag{2.33}$$

$$\text{sgn}(x) = \begin{cases} 1, & x \geq 0 \\ -1, & x < 0 \end{cases} \tag{2.34}$$

where r is the range, θ is the incidence angle, H is the satellite altitude, and H_s is the storm height. This approximation is valid under the condition that $c\tau \ll 2r\theta_B$; (i.e., the range resolution is narrow relative to the field of view).

Figure 2.3 illustrates the variation in fractional fill parameter, F, as a function of range for a satellite altitude of 350 km, a storm height of 10 km, and an incidence angle of 30° for several values of the radar beamwidth, θ_B. More general examples of partial beamfilling and their effect on various rain rate retrieval methods have been investigated [6–7].

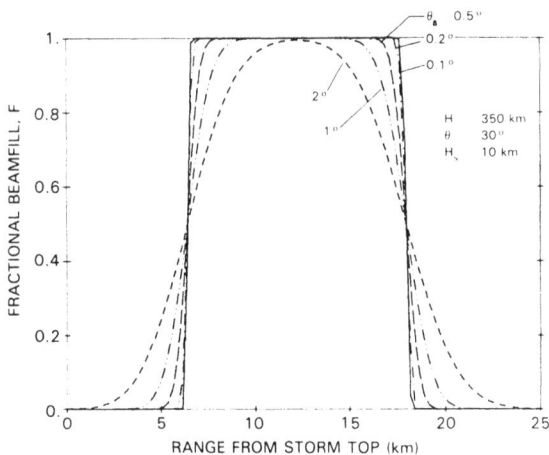

Figure 2.3 illustrates...

Figure 2.3 (chart: FRACTIONAL BEAMFILL, F vs RANGE FROM STORM TOP (km); curves labeled θ_a 0.5°, 0.2°, 0.1°, 2°, 1°; H 350 km, θ 30°, H_s 10 km)

Figure 2.3 Gain-weighted fractional filling of the pulse volume as a function of range from the storm top for several values of the 3 dB antenna beamwidth.

Receiver Losses

One type of processing loss arises from the finite bandwidth of the receiver [8–9]. In the derivation of the radar equation, the envelope of the signal at the receiver input, $u(t)$, was assumed to be rectangular. More generally, the integration should be performed on the output of the receiver filter. Following Doviak and Zrnić [9], the range integration can be expressed as a product of the output from an infinite receiver bandwidth, $c\tau/2$, and a factor L:

$$\int |v(t - 2r/c)|^2 \, dr = Lc\tau/2 \tag{2.35}$$

where $v(t)$ is the complex envelope at the receiver filter output. For a rectangular input pulse of duration τ and a filter having a Gaussian frequency response given by $\exp(-4 \ln 2 \, f^2/B^2)$ where B is the 6 dB bandwidth, a good approximation for $B\tau \gg 1$ is [9]:

$$L = \coth(aB\tau) - 1/aB\tau; \quad a = \pi/(2 \sqrt{\ln 2}) \tag{2.36}$$

The factor L increases monotonically with $B\tau$, approaching unity as $B\tau \to \infty$. For $B\tau = 1$, a numerical integration of (2.35) yields $10 \log L = -2.3$, while for a matched filter $10 \log L = -1.7$.

Spatial Resolution

Because the return power is proportional to the square of the gain pattern, the angular resolution can be defined as the locus of points of the two-way gain pattern that are 6 dB down from the point of maximum return. For a circular Gaussian pattern, the fraction of the total power from uniformly distributed targets within this contour is approximately $\Phi(\sqrt{2 \ln 2})$, or about 90%, where the error function Φ is defined by (2.33). (We note that in comparing the radar resolution with a passive sensor such as a microwave radiometer, where the gain pattern enters into the equation for the brightness temperature only on reception, the angular resolution of the two instruments is matched when the radar beamwidth is equal to $\sqrt{2}$ times that of the radiometer.) For the range resolution, an analogous definition can be used. Letting v be the complex envelope of the signal at the receiver filter output, the normalized output power can be written as

$$P(r) = |v(r)|^2 / (|v(r)|_{max})^2 \tag{2.37}$$

Denoting the two solutions of $10 \log P(r) = -6$ by r_1 and r_2, the range resolution, h, can be defined as $h = |r_2 - r_1|$. Assuming that the received signal at the antenna input is a rectangular pulse of duration τ, and that the filter frequency response is a Gaussian of 6 dB bandwidth B, h can be approximated by [9]:

$$h = (1/aB\tau)(0.5 \; c\tau) \cosh^{-1}[2 + \cosh(aB\tau)]; \; a = \pi/(2 \sqrt{\ln 2}) \tag{2.38}$$

For $B\tau = 1$, $h = 1.17 \; c\tau/2$, while for $B\tau \to \infty$, h approaches the standard result $c\tau/2$.

Effective Vertical Resolution

For a spaceborne weather radar, the resolutions of interest generally do not correspond with the alignment of the pulse volume. The definition of an effective vertical resolution is useful in assessing the degradation of the vertical resolution when the beam is scanned away from nadir. As shown in Figure 2.4, the vertical extent of the pulse volume is comprised of the part due to the range resolution, h_r, and the part due to the finite beamwidth h_b where $h_r = h \cos \theta$ and $h_b = r\theta_B \sin \theta$. A definition of the effective vertical resolution can be obtained by considering the

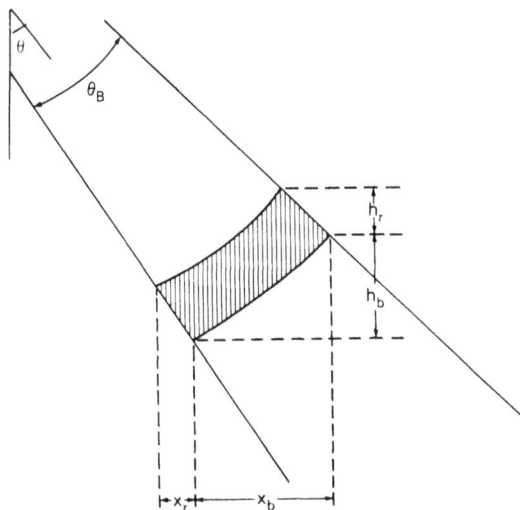

Figure 2.4 Geometry of the pulse volume for off-nadir incidence.

average response function for the received power along the vertical direction z [10]:

$$\bar{P}(z) = \iint P(x,y,z)\, \mathrm{d}x\, \mathrm{d}y \tag{2.39}$$

where $P(x,y,z)$ is obtained from a coordinate transformation from the local coordinates centered on the pulse volume to the horizontal (x,y) and vertical (z) frame of reference. Denoting the two solutions of $10\log\bar{P}(z) = -6$ by z_1 and z_2, then the effective vertical resolution, h_e, is defined as $h_e = |z_2 - z_1|$. If $h_b > h_r$, a good approximation to h_e has been shown to be [10]:

$$h_{RSS} = (h_r^2 + h_b^2)^{1/2} \tag{2.40}$$

which, by the triangle inequality, is always less than the maximum vertical extent of the pulse volume, h_{P-P}, given by the sum of h_r and h_b. Comparisons of h_{P-P}, h_e, and h_{RSS} are illustrated in Figure 2.5 as a function of incidence angle. Similar definitions can be used to define the effective horizontal resolution. For example, from Figure 2.4, the effective resolution along the x direction is approximately $(x_b^2 + x_r^2)^{1/2}$.

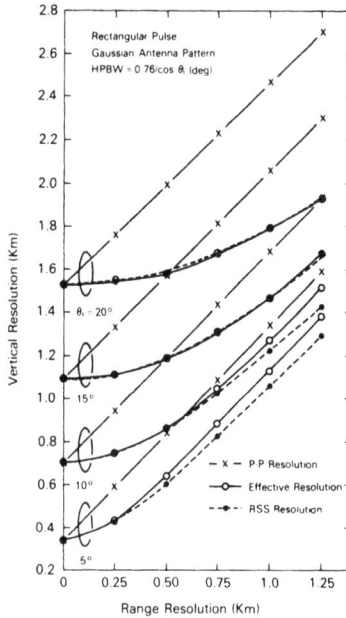

Figure 2.5 Vertical resolutions *versus* range resolution for four incidence angles (after Kozu, [10]).

2.2 SURFACE RETURN

For nadir incidence and a narrow transmitted pulse, the backscattered power from the hydrometeors arrives at the receiver prior to the large surface contribution. At off-nadir angles, however, the surface return arrives simultaneously with this return, and increases in magnitude as the range gates within the main beam begin to intersect the surface. Thus, the surface clutter can mask one of the quantities of greatest interest in hydrological and agricultural applications: the rainfall rate at the surface. Therefore, one of the design requirements, is to restrict the influence of the clutter to as small a region as possible. This requires a narrow main beam in elevation and low sidelobe levels; in particular, if the main beam is directed along the elevation angle θ_0, it is necessary to minimize the sidelobe levels for elevation angles less than θ_0. For radars that employ pulse compression techniques, low range (or time) sidelobe levels are needed as well.

Main Beam Surface Contribution

For our purposes, an equation for the surface return power, P_s, can be found by replacing $\eta \, dV$ in the meteorological radar equation with $\sigma^\circ \, dS$ where σ° is the normalized (or differential) backscattering cross section of the surface (area/area) and dS is an area element:

$$P_s(r) = (P_0\lambda^2/(4\pi)^3) \int_S (\sigma^\circ G^2 |u(t - 2r/c)|^2 e^{-0.2 \ln 2 \int_0^r k \, ds/r^4}) \, dS \tag{2.41}$$

In the following discussion we assume that $|u(t)| = 1$ for $\tau > t > 0$, and zero elsewhere. The region of integration in (2.41) can be visualized as the intersection of the earth's surface with an expanding spherical shell of thickness $c\tau/2$ (Figure 2.6(a)). For times less than $2H/c$ (where H is the satellite altitude, c is the speed of light and $t = 0$ is associated with the leading edge of the pulse at the transmitter), $P_s = 0$. For $2H/c + \tau > t > 2H/c$, the area of integration is a circle of radius ρ_0, where (flat-earth approximation):

$$\rho_0 = [(ct/2)^2 - H^2]^{1/2} \tag{2.42}$$

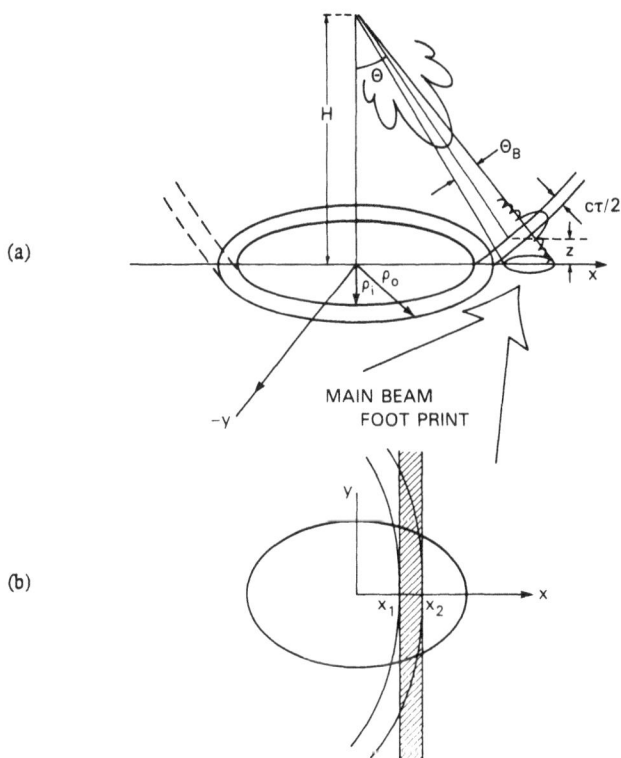

Figure 2.6 (a) surface return from an annular ring; (b) surface scattering from the mainlobe, approximating the annular integration with a strip parallel to the y-axis.

For $t > 2H/c + \tau$, the surface return arises from an annulus of outer radius ρ_0, as given above, and inner radius ρ_i, where

$$\rho_i = [(c(t - \tau)/2)^2 - H^2]^{1/2} \tag{2.43}$$

An important special case occurs when: (1) ρ_i is much greater than the FOV; (when $\rho_i \gg r\theta_B$, where r is the radar range and θ_B is the 3 dB elevation beamwidth) and (2) where the main beam intersects a portion of the annulus. Using the Gaussian pencil-beam approximation for G and replacing the annulus with a strip along the y-axis (Figure 2.6(b)) gives

$$P_s(r) = \frac{P_0\sigma^\circ(G_0\lambda)^2\theta_B\phi_B e^{-0.2\ln10 \int_0^r k\,ds}}{2^{10}\pi^2\ln2\ r^2\cos\theta} [\operatorname{sgn}(x_2)\Phi(q|x_2|) - \operatorname{sgn}(x_1)\Phi(q|x_1|)] \tag{2.44}$$

where the functions Φ and sgn are defined by (2.33) and (2.34), respectively, and

$$q = \sqrt{8\ln2}\ \cos^2\theta/H\theta_B \tag{2.45}$$

$$x_1 = \rho_i - H\tan\theta \tag{2.46}$$

$$x_2 = \rho_0 - H\tan\theta \tag{2.47}$$

Recall that θ_B, ϕ_B are the elevation and azimuthal beamwidths, respectively, and θ is the off-nadir incidence angle.

For $qx_2 \gg 1$ and $q|x_1| \gg 1(x_1 < 0)$, the term in brackets approaches 2. This condition is referred to as *beam-limited*, as the area of integration is determined by the extent of the beamwidth. The resulting P_s holds for all incidence angles. The other extreme occurs when the width of the annulus (the projection of the range resolution onto the surface) is small relative to the linear footprint dimension $r\theta_B$. For this *pulse-limited case*, the bracketed term in (2.44) is approximately

$$(c\tau q/\sqrt{\pi}\sin\theta)\exp\{-2\ln2[(x_1 + x_2)\cos^2\theta/H\theta_B]^2\} \tag{2.48}$$

In contrast to the beam-limited case, the approximation is valid only at angles for which $\rho_i \gg r\theta_B$.

In the special case when the range gate intercepts the center of the FOV, then $x_2 = -x_1$, and the surface return attains its maximum value. (An exceptional case occurs if the attenuation is very large, in which case the surface maximum will occur earlier). Accounting for the fact that half the pulse volume is filled with rain, then the ratio of the rain (2.19) to the surface return is

$$P_r/P_s = c\tau\eta\cos\theta/(4\sigma^\circ\Phi(qx_2)) \tag{2.49}$$

where $x_2 = c\tau/(4\sin\theta)$.

For the beam-limited case, $\Phi(qx_2) \to 1$, so the ratio is directly proportional to the product of the rain reflectivity η and the range resolution, and inversely proportional to the backscattering cross section. For most surfaces, σ° decreases with incidence angle more rapidly than cos θ, so that P_r/P_s is an increasing function of incidence angle. Although the frequency dependence of P_r/P_s is difficult to specify, in most cases the ratio increases with frequency. In the case of agricultural crops, the data of Ulaby [11] suggest that at nadir incidence, σ° increases by about 2.5 dB as frequency increases from 10 GHz to 18 GHz, whereas at incidence angles of 20° and 60°, σ° increases by about 4 dB over the same span of frequencies. For rain, however, assuming Rayleigh scattering, η is proportional to the fourth power of frequency. These relationships imply that for incidence angles between 20° and 60°, P_r/P_s will increase by about 6 dB when the radar frequency is raised from 10 GHz to 18 GHz; at nadir the increase is expected to be about 7.5 dB.

For the pulse-limited case (when the range gate intercepts the center of the FOV), the ratio of the rain to the surface return exhibits the same frequency dependence as in the beam-limited case but grows more rapidly with increases in the off-nadir incidence angle:

$$(P_r/P_s) = \sqrt{\pi} \, \eta H \theta_B \tan \theta / (4 \sqrt{2 \ln 2} \, \sigma^\circ) \tag{2.50}$$

In contrast to the beam-limited result, the ratio is proportional to the beamwidth. To increase the rain return by broadening the beam, however, exacts a price in a loss in vertical resolution and an increase in the number of range gates that are partially filled with rain.

Figure 2.7(a) shows curves of (P_r/P_s) (dB) *versus* incidence angle for five values of rain rate. The σ° data were approximated from the Skylab scatterometer results over the ocean [12] ($f = 13.9$ GHz). For both figures, the beamwidth is taken to be 0.72° and $c\tau/2 = 250$ m, which are parameters of the proposed TRMM radar. Because of the high degree of sensitivity of σ° to windspeed (especially for angles greater than about 20° [13]), we should understand that a high degree of variability exists in the (P_r/P_s) curves. Using values from the vegetation clutter model of Ulaby [11], Figure 2.7(b) shows curves of P_r/P_s (dB) *versus* frequency for five values of incidence angle at a rainfall rate of 5 mm/h. For the calculation of P_r, Rayleigh scattering has been assumed throughout.

Influence of the Surface on Pulse Compression

Although pulse compression has a number of advantages (see Section 2.5), one serious drawback for meteorological sensing is the masking of the rain signal caused by the intersection of the range (or time) sidelobes with the surface. To treat this case, we let the central pulse volume be located a distance z above the

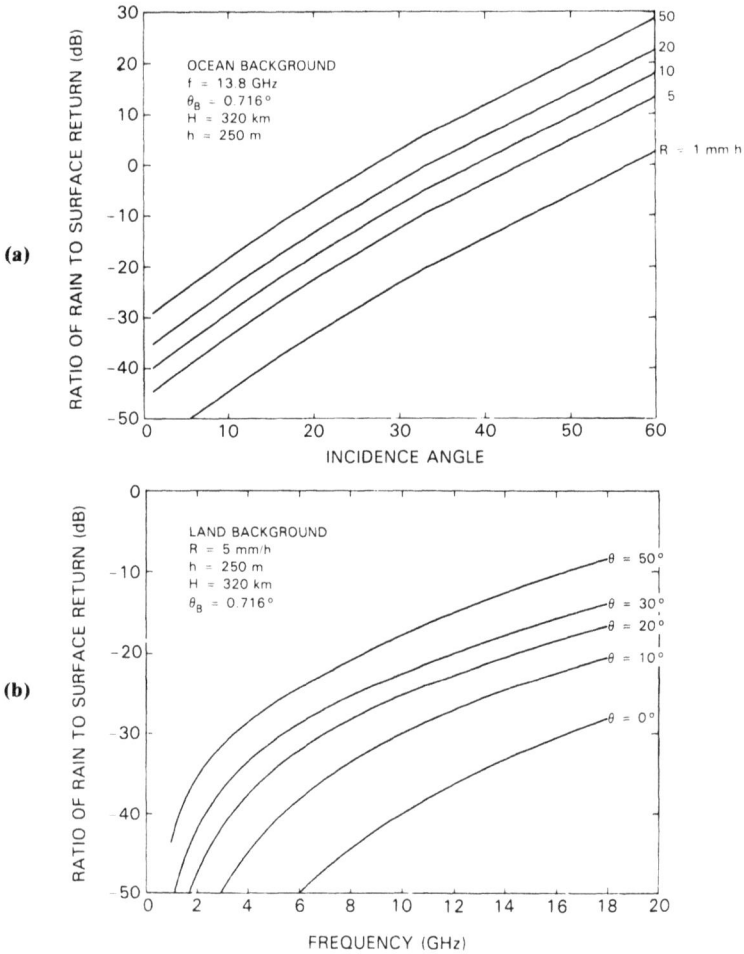

Figure 2.7 (a) Ratio of the rain to the surface return (dB) over ocean *versus* incidence angle for several rain rates; (b) Ratio of the rain to the surface return (dB) over land *versus* frequency for several incidence angles at a rain rate of 5 mm/h.

surface and assume that the range sidelobe level that intercepts the surface is reduced by a factor F relative to the level within the central pulse volume (Figure 2.6(a)). To obtain an expression for P_r/P_s for this case, we first modify (49) by $2/F$, where the factor of two corresponds to a pulse volume fully filled with rain, and by a factor to account for the different locations of the surface and rain returns. The ratio of the rain to the surface return becomes

$$\frac{P_r}{P_s} = \frac{c\tau\eta \cos \theta H^2}{2\sigma°F(H-z)^2} \exp \left(0.2 \ln 10 \int_{r_1}^{r_2} k \, ds\right) \tag{2.51}$$

where

$$r_1 = (H - z) \sec \theta; \; r_2 = H \sec \theta \qquad (2.52)$$

In Figure 2.8, contour plots of (P_r/P_s) (dB) are shown as functions of altitude and rainfall rate for $10 \log F = -50$ [14]. At nadir incidence for a radar frequency

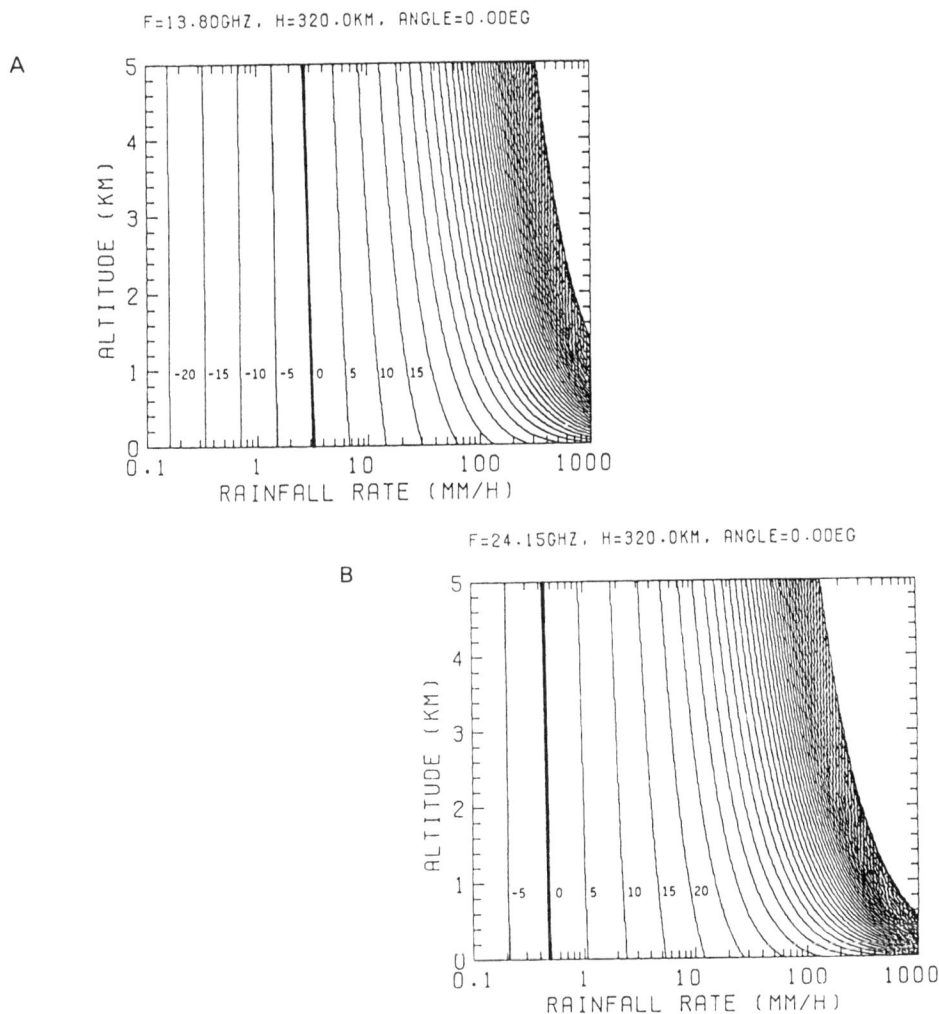

F=13.80GHZ, H=320.0KM, ANGLE=0.0DEG

A

F=24.15GHZ, H=320.0KM, ANGLE=0.0DEG

B

Figure 2.8 Contour plots of the ratio of the rain to the surface return for range sidelobe levels down by 50 dB: (a) $f = 13.8$ GHz, nadir; (b) $f = 24.15$ GHz, nadir; (c) $f = 13.8$ GHz, $20°$ (from Manabe and Ihara [14]).

F=13.80GHZ, H=320.0KM, ANGLE=20.00EG

C

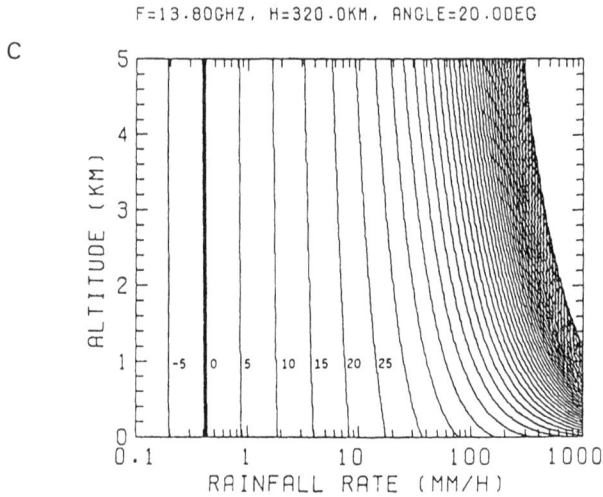

Fig. 2.8 continued

of 13.8 GHz, the surface contribution from the range sidelobes exceeds the rain return for rainfall rates below about 3 mm/h (Figure 2.8a). To attain a 10 dB signal-to-clutter margin at the surface, the rainfall rate must exceed 10 mm/h. When the incidence angle or the frequency is increased, the clutter is reduced. Results for a 24 GHz frequency at nadir incidence and a 13.8 GHz frequency at an angle of 20°, are shown in Figures 2.8(b) and 2.8(c), respectively. In both cases, the rain rate above which P_r exceeds P_s is well below 1 mm/h. Although the technology is expected to make possible achieving range sidelobe levels below 55 dB by the mid-1990s, even with such performance, some difficulties will be present at lower frequencies and at near-nadir angles.

Sidelobe Surface Clutter

To compare the rain return to the surface contributions from the main and antenna sidelobes, the general expressions for P_r and P_s can be integrated numerically. In Figures 2.9 and 2.10, a simple stratiform model is used consisting of snow (1.5 km), mixed phase hydrometeors or melting layer (0.6 km), and a rain layer (5 km), so that the total storm height is 7.1 km. The radar parameters are approximately the same as the TRMM radar (see Section 3.7). Values of $\sigma°$ are appropriate to moderate windspeeds over the ocean [15]. The sets of curves in each figure correspond to radar return powers from the hydrometeors and the surface as a function of the radar range measured from the storm top. To emphasize the clutter contribution from the sidelobes, a circular aperture with uniform illumination is selected, so that the level of the first sidelobe (one-way pattern) is down 17 dB from the on-axis gain. For an incidence angle of 10°, results are shown for light (Figure 2.9(a)) and moderate (Figure 2.9(b)) rainfall rates. To reduce the extent of the region where the clutter masks the rain, lower sidelobes are needed.

At an incidence angle of 30°, Figure 2.10(a) shows that despite negligible sidelobe levels even for light rainfall rates, the surface return along the main lobe

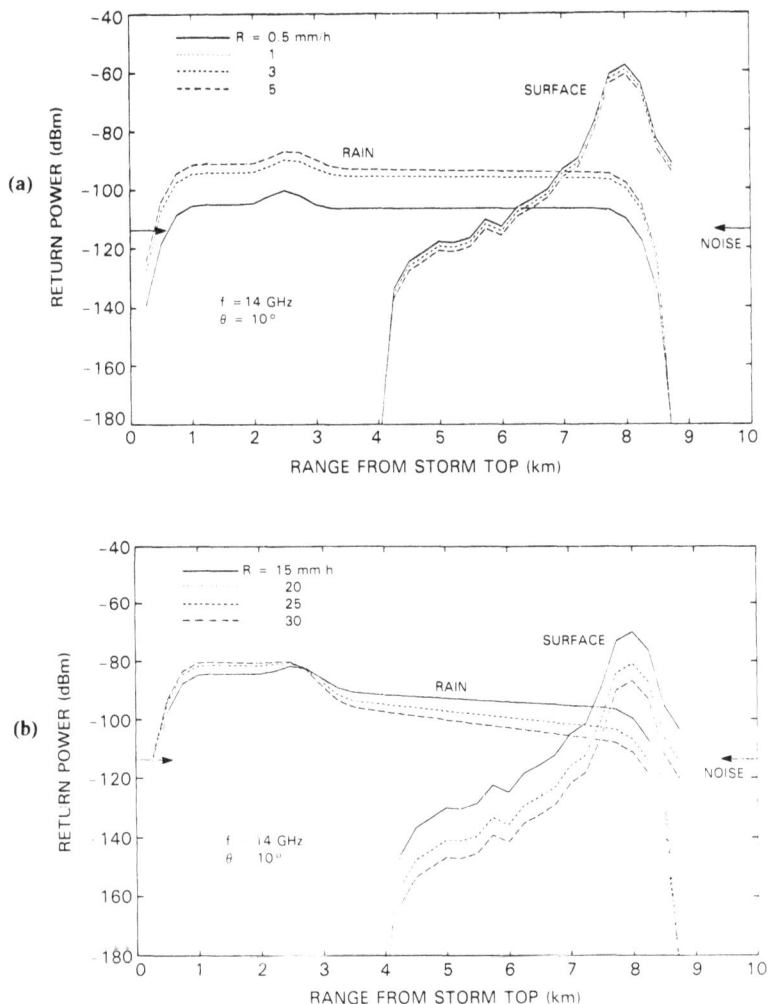

Figure 2.9 Return powers from the rain and surface over ocean at an incidence angle of 10° and $f = 14$ GHz for (a) light rain rates and (b) moderate rain rates [courtesy of J.A. Jones].

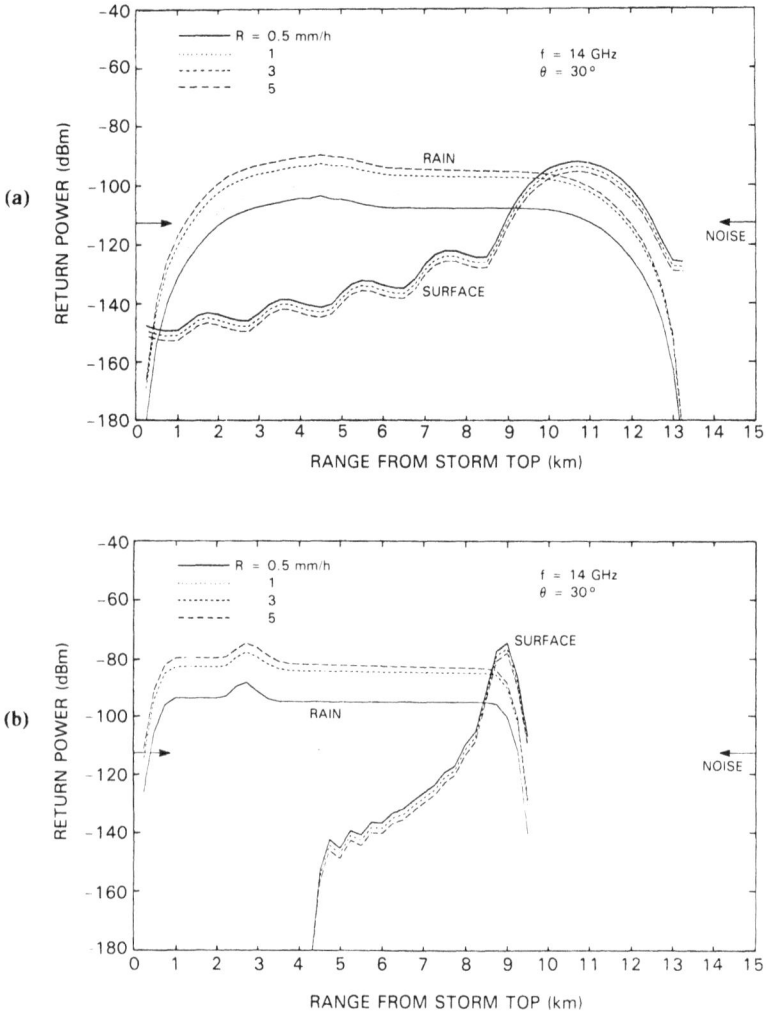

Figure 2.10 (a) Same as Figure 2.9, but for an incidence angle of 30°; (b) Same as (a) but for a narrow beamwidth (D/λ = 400) [courtesy of J.A. Jones].

obscures the rain signal over a wide range. Further restricting the influence of the surface clutter requires a narrower beamwidth. Figure 2.10(b) shows an example of the rain and surface returns at θ = 30° for an antenna diameter of 400 wavelengths. Not only does the larger antenna restrict the surface effects, but it provides enhanced definition of the storm echo top and the melting layer.

2.3 MIRROR-IMAGE RETURN

Airborne meteorological radars have observed not only the returns from the hydrometeors and the surface, but also contributions that arrive subsequent to them. These contributions include the backscattered surface power transmitted and received along the antenna sidelobes, and surface energy bistatically scattered out to a volume of rain which scatters a portion of the signal toward the receiving antenna. At near-nadir incidence over the ocean, the dominant component seems to arise from the scattering of the incident energy from the surface out to the rain, a fraction of which is returned to the surface, where the signal is scattered a second time back to the radar [16–18]. This doubly-reflected surface return has been denoted the *mirror-image (MI) return*. A schematic of one such scattering path is shown in Figure 2.11.

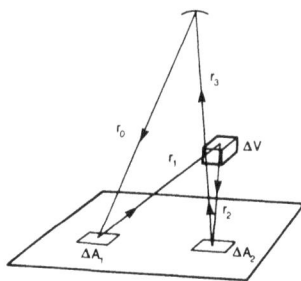

Figure 2.11 Schematic of a scattering path that contributes to the mirror-image return (from Meneghini and Atlas [17]).

Examples of measured radar returns obtained from an airborne radar at 10 GHz are indicated by the heavy solid lines in Figure 2.12. The scale along the ordinate is 10 log of the product of the radar return power and the square of the radar range. In these cases, where the beam is pointed at nadir over an ocean background, the large, narrow surface return (located about 5 km below the radar in Figure 2.12(a)) clearly separates the direct rain return on the left from the mirror-image return on the right. We note that in the example of stratiform rain (Figure 2.12(b)) (which shows a well defined melting layer at a distance of about 1 km from the radar), the degree of similarity between the direct and mirror-image components is higher than the convective rain shown in Figure 2.12(a).

For spaceborne weather radars, the mirror-image component may influence the choice of the *pulse repetition frequency* (PRF). One of the constraints in the selection of the PRF is that not more than one pulse may be present in the scattering medium at any given time. If the mirror-image return is significant, then the effective length of the scattering medium is increased. Consequently, the upper bound on the PRF decreases, so that fewer pulses are allowed per scan. For

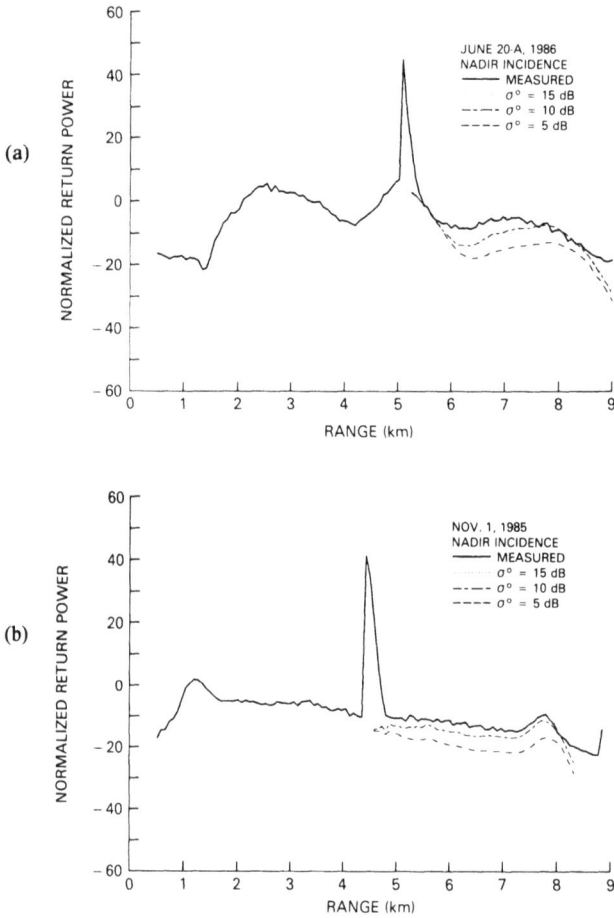

Figure 2.12 Range normalized radar return power at $f = 10$ GHz *versus* range from the aircraft. Comparisons of the measured mirror image return with theoretical values for (a) convective rain and (b) stratiform rain (after Meneghini and Nakamura [18]).

some radar designs, this will limit the maximum allowable swath width (see Section 3.4).

To get an idea of the magnitude of the mirror-image component, an approximation for the ratio of this component, $P_m(t)$ to the direct rain return, $P_d(t')$ for a range gate a distance z above the surface is [17]:

$$\frac{P_m(t)}{P_d(t')} - \frac{(H - z)^2 \Gamma^4 \sigma° e^{-0.4 \ln 10 \int_0^z k \, dz}}{\sigma° H^2 + p \Gamma^2 z^2 / \theta_B^2} \qquad (2.53)$$

The times t and t' are related by $t = t' + 4z/c$, Γ^2 is the Fresnel reflectivity of the surface, H is the radar altitude and the constant p is approximately equal to 11. At ranges close to the surface, (2.53) predicts that the ratio is equal to Γ^4. Over an ocean background at nadir, Γ^2 is approximately 0.63, implying that the mirror-image component just "below" the surface is about 4 dB less than the rain return just above the surface. Experimental data (and numerical simulations) suggest that the difference is somewhat smaller than 4 dB. In fact, because of the number of assumptions used in deriving $P_m(t)$, caution is needed in its application. Comparisons of measured (solid curves) and predicted MI returns (discontinuous curves) are shown in Figure 2.12, using a somewhat more general expression than (2.53) [18]. Over land at near-nadir incidence angles, the MI return will generally be much smaller; for example, if $\Gamma^2 = 0.2$, the MI return will be 14 dB smaller than the rain return.

Apart from the magnitude of $P_m(t)$, one other question of interest is the volume of rain to which $P_m(t)$ corresponds. Comparisons of the effective MI rain volumes are shown in Figure 2.13 for spaceborne and airborne weather radar geometries for $\sigma^\circ = 10$ dB. Contours of -3 dB, -6 dB, and -9 dB have been plotted for three heights (observation times). The levels have been defined relative to the maximum value so that the -3-dB contour is given as the set of points which are down 3 dB from the point at which the MI return is maximum. In each figure, the 6-dB antenna beamwidth (two-way pattern) is shown. We can infer from these results that part of the reason for the distortion in the image return relative to the direct return is due to the broadening of the scattering volume caused by the free-space divergence of the beam, as well as broadening caused by the surface. The broadening increases for rougher surfaces (lower σ° values) and for larger beamwidths. For a "typical" narrow-beam spaceborne radar, the direct and mirror-image rain volumes will be better matched than for most airborne radars, as we can see by comparing Figures 2.13(a) and 2.13(b). Matched MI and direct beams should improve the capability of estimating the path attenuation (and therefore rainfall rate) from P_m and P_d in much the same way as matched beams at two frequencies improve the accuracy of the attenuation estimate (see Section 5.3).

2.4 SYSTEM NOISE POWER

The *signal-to-noise ratio* (S/N) and the number of independent samples are the two major factors which determine the accuracy of the estimate of signal power (see Section 2.5). In this section, we describe some basic properties of the receiver noise. The S/N which affects the estimation accuracy is that at the final stage of the receiver. Although some degradation of S/N may arise from such

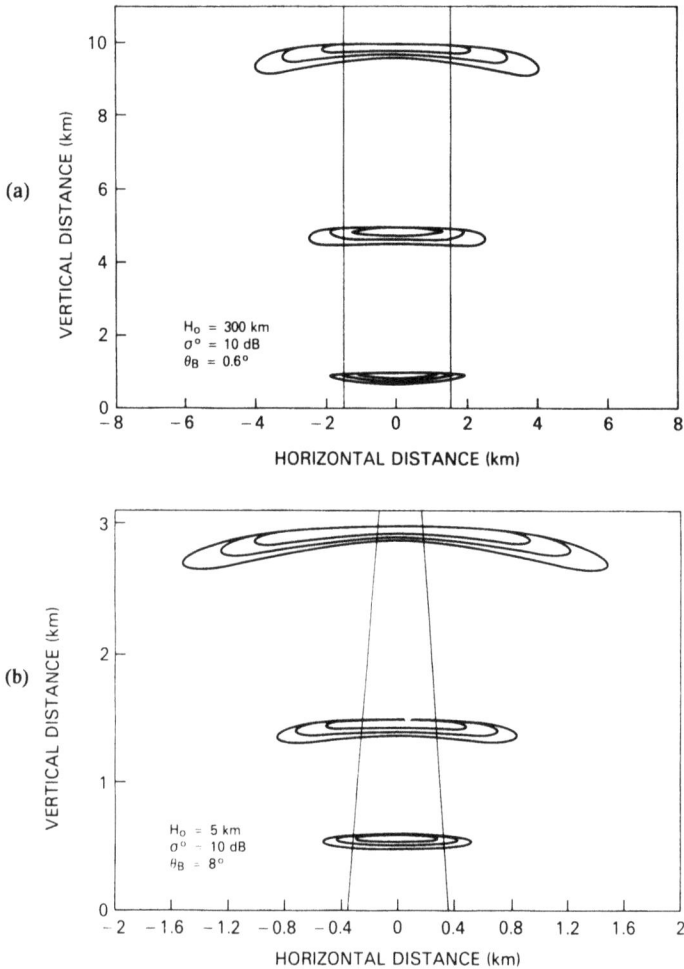

Figure 2.13 Effective rain volumes that contribute to the mirror image return for (a) a spaceborne radar geometry and (b) an airborne geometry (after Meneghini and Nakamura [18]).

factors as quantization noise in the later stages of the receiver, these are usually negligible relative to the thermal noise at the antenna output and the noise generated in the receiver RF and IF stages.

The noise power per unit bandwidth at the receiver output, N, is usually characterized by the system noise temperature, T_{sys} via the relationship:

$$N = kT_{sys}G \tag{2.54}$$

where k is the Boltzmann's constant $(1.38 \times 10^{-23}$ (W s)/K), and G is the total receiver gain. If the receiver is noise-free, T_{sys} is equal to the antenna noise temperature T_a, which is defined by

$$N_a = kT_a \tag{2.55}$$

where N_a is the noise power per unit bandwidth at the antenna output. In practice, however, the receiver generates additional noise, so that T_{sys} is larger than T_a. We may express this degradation by

$$T_{\text{sys}} = T_a + T_e \tag{2.56}$$

where T_e is the effective receiver noise temperature. The same definitions can be applied to each receiver component.

The receiver usually consists of a series of amplifiers, attenuators, and mixers. The relation between the "total" effective noise temperature (T_{rec}) and that of each component is given by

$$T_{\text{rec}} = T_1 + T_2/G_1 + T_3/(G_1G_2) + \cdots + T_n/(G_1G_2 \cdots G_{n-1}) \tag{2.57}$$

where $T_1, T_2, \ldots T_n$ are the effective noise temperatures of the front-end, 2nd stage, and n-th stage components, respectively, and where $G_1, G_2, \ldots G_{n-1}$ are the gains of the corresponding components. Note that G has a value less than unity for an attenuator or mixer. As shown in (2.57), the contribution to T_{rec} from the noise temperatures of second and later stages can be made negligible by choosing a front-end amplifier gain that is sufficiently large.

The other way to characterize the receiver noise performance is by the noise figure, F, which is defined as the S/N at the receiver (or component) output to that at the input, where the input is assumed to be terminated by a matched load at temperature T_0. By convention, T_0 is taken to be 290 K. The quantities F and T_e are related by

$$F = (T_0 + T_e)/T_0 \tag{2.58}$$

The relationship between the total receiver noise figure, F_{rec}, and the noise figures of the various components, F_i, $(i = 1, \ldots n)$ is given by

$$F_{\text{rec}} = F_1 + (F_2 - 1)/G_1 + (F_3 - 1)/(G_1G_2) + \cdots + (F_{n-1})/(G_1G_2 \cdots G_{n-1}) \tag{2.59}$$

As an example, consider the effective noise temperature and noise figure of an attenuator at temperature T. The gain of the attenuator is given by $1/L$ where L is the loss factor. Because the noise temperature due to the attenuator is $(1 - 1/L)T$ and the effective noise temperature is defined at the input of the attenuator, then,

$$T_{e(att)} = L(1 - 1/L)T = (L - 1)T \tag{2.60}$$

Therefore, the noise figure is

$$F_{(att)} = (T_0 + T_{e(att)})/T_0 = 1 + (L - 1)T/T_0 \tag{2.61}$$

If $T = T_0$, $F_{(att)}$ is equal to L. As the noise figure of the front-end contributes most to the total noise figure, it is crucial to reduce the feed loss between antenna and receiver to avoid degradation of receiver noise performance, especially in the case of low-noise receivers. Assuming a receiver front-end in which a feed line (loss factor L and temperature T_0) and an amplifier (noise figure F and gain G) are connected in series, the total noise figure F_{rec1} is

$$F_{rec1} = L + (F - 1)L = FL. \tag{2.62}$$

On the other hand, if the connections are reversed,

$$F_{rec2} = F + (L - 1)/G \tag{2.63}$$

In the former case, the effect of L on the total noise figure may be significant. For the latter case, the second term of (2.63) is often negligible so that F_{rec2} is approximately equal to F.

The above noise figure is defined only for receiver noise performance. A similar quantity called *system noise factor* or *operating noise factor* (F_s) is sometimes used for the system (receiver plus antenna noise) performance [19]. This is defined by

$$F_s = T_{sys}/T_0 = T_a/T_0 + F - 1 \tag{2.64}$$

The antenna noise temperature T_a, defined by (2.55), consists of two factors: the noise temperature entering the antenna, T_{a0}, which is equal to the antenna noise temperature without ohmic losses, and the noise generated by ohmic losses in the antenna, L_a. Note that T_{a0} is defined as the product of the upwelling brightness temperature and the normalized antenna gain pattern integrated over all space [12]. The resulting T_a is given by

$$T_a = T_{a0}/L_a + (1 - 1/L_a)T_0 \tag{2.65}$$

If the various parts of the antenna have different physical temperatures, the second term of (2.65) should be expanded into component form.

For a low noise receiver, T_{a0} can contribute substantially to T_{sys}. As the use of microwave radiometry for remote sensing of the earth and the atmosphere makes clear, T_{a0} varies significantly with the surface (land/ocean) and with the constituents of the atmosphere. Therefore, some caution must be exercised in the use of algorithms that attempt to subtract the noise component from a measurement of signal plus noise if the measurements are made under different atmospheric or surface conditions.

2.5 STATISTICAL CONSIDERATIONS

The complex voltage at the input to the receiver can be expressed in the form:

$$v = \sum_i A_i \exp(-i\psi_i) \tag{2.66}$$

where A_i, ψ_i are the amplitude and phase of the scattered field from the ith particle. For a monostatic radar, $\psi_i = 2[kr - \pi(f_0 + f_{Di})t]$ where f_0 is the radar frequency, f_{Di} is the doppler frequency shift from the ith particle and k is the wavenumber $(2\pi/\lambda)$. If the radar resolution is large with respect to the wavelength and the scatterers are large in number and randomly distributed, then the phase of v is uniformly distributed over 0 to 2π and the amplitude follows the standard Rayleigh distribution. Equivalently, the real and imaginary parts of v (denoted by v_I and v_Q, respectively) can be modeled as independent Gaussian random variables of zero mean each with a variance σ^2 proportional to the sum of the squares of the individual amplitudes, A_i [19]. Using the notation $E(P) = \bar{P}$ for the mean value of the power, P, where $P = vv^* = v_I^2 + v_Q^2$, then,

$$\bar{P} = 2E(v_I^2) = 2E(v_Q^2) = 2\sigma^2 = \sum_i |A_i|^2 \tag{2.67}$$

For the rain, surface, and system noise, $\bar{P}$ is associated with the corresponding expressions for P_r, P_s and P_N, respectively.

The sum of squares of two identical zero-mean Gaussian random variables follows an exponential *probability density function* (pdf) so that the pdf for the power P can be written as

$$p(P) = (1/\bar{P}) \exp(-P/\bar{P}) \tag{2.68}$$

from which it follows that the standard deviation of P is equal to the mean value, $\bar{P}$. To reduce the variance of an estimate based on a single sample, the estimate for P, denoted by $\hat{P}$, is taken to be the sample mean of N independent random variables, each with a pdf given by (2.68). The result can be written in the form [21]:

$$\hat{P} = \bar{P}f_N \tag{2.69}$$

where the pdf of f_N is

$$p(f_N) = N^N f_N^{N-1} \exp(-Nf_N)/(N - 1)! \tag{2.70}$$

so that $E(f_N) = 1$ and $\text{var}(f_N) = 1/N$. Therefore, the estimate of P is unbiased, $E(\hat{P}) = \bar{P}$, with a variance inversely proportional to the number of independent samples: $\text{var}(P) = (\bar{P})^2/N$.

The mean and variance of the estimate given above can be used even if the samples are partially correlated as long as N is replaced with an effective number of independent samples, N_e. Defining N_e as the square of the mean of P to the variance of P and noting that [22]:

$$\text{var}(\hat{P}) = ((\bar{P})^2/N) \left[1 + (2/N) \sum_{k=1}^{N-1} (N - k)\rho(kT_p) \right] \tag{2.71}$$

where T_p is the interpulse period and $\rho(kT_p)$ is the correlation coefficient for samples spaced kT_p apart, then,

$$N_e = N^2 \Big/ \left[N + 2 \sum_{k=1}^{N-1} (N - k)\rho(kT_p) \right] \tag{2.72}$$

Note that $N_e \leq N$, where the equality holds if the correlation coefficients are zero.

The above results are based on the assumption that the video output of the receiver is proportional to the backscattered power (square-law detector). However, due to the large dynamic range of meteorological targets, a logarithmic detector is used in most conventional (incoherent) radars. If N samples of the voltage are logarithmically detected and then summed without first taking the antilog, then the estimate of the return power is

$$\hat{P}_{\log} = \exp \left[(2/N) \sum_{i=1}^{N} \ln(|V_i|) \right] \tag{2.73}$$

where the random variable $|V_i|$, the envelope of the voltage at the intermediate frequency stage, is Rayleigh distributed.

For a linear detector, the power estimate is

$$\hat{P}_{\text{lin}} = \left[(1/N) \sum_{i=1}^{N} |V_i| \right]^2 \tag{2.74}$$

For N greater than about 30, the mean of these estimates is independent of N and given by [9,20]:

$$E(\hat{P}_{\log}) = 0.56 \, \overline{P} \tag{2.75}$$

$$E(\hat{P}_{\text{lin}}) = \pi \overline{P}/4 \tag{2.76}$$

which constitutes a bias of -2.5 dB for the logarithmic detector and a -1 dB bias for the linear detector. To account for these processing losses, a multiplicative factor of 0.56 (logarithmic detector) or 0.79 (linear detector) should be introduced into the right-hand side of the various radar equations.

If the number of independent samples is greater than about 10, the variances of the unbiased estimates, $\hat{P}'_{\log} = \hat{P}_{\log}/0.56$, $\hat{P}'_{\text{lin}} = 4\hat{P}_{\text{lin}}/\pi$, are given by [9]:

$$\text{var}(\hat{P}'_{\log}) = (\pi \overline{P})^2/6N \tag{2.77}$$

$$\text{var}(\hat{P}'_{\text{lin}}) = 1.1(\overline{P})^2/N \tag{2.78}$$

which are slightly larger than in the case of square law detection: $\text{var}(P) = (\overline{P})^2/N$. Expressions for the effective number of independent samples when the samples are partially correlated are given by Walker et al. [23] and Doviak and Zrnic [9].

Decorrelation Effects

The severe power constraints of spaceborne radar force a choice of a sufficiently large T_p so that $\rho(kT_p)$ is small for all k. (For the case analyzed here, $\rho(t)$ is a monotonically decreasing function of t, so the above condition is equivalent to requiring that $\rho(T_p)$ is small.) This condition ensures that the samples are virtually uncorrelated, and therefore are used efficiently to reduce the error variance of the power estimate. In Section 2.8, we return to this question from the perspective of doppler measurements.

Although a number of processes act to decorrelate successive return pulses, for most spaceborne geometries, the largest source of decorrelation is spacecraft motion. One way to compute this is to let $v_1 = \Sigma A_i \exp(-i\psi_i)$ and $v_2 = \Sigma A_i \exp(-i\phi_i)$ be proportional to the fields returned from an ensemble of scatterers when the antenna is located at $(-L/2, 0, H)$ and $(L/2, 0, H)$, respectively, where H is the altitude of the spacecraft traveling at a speed of V along the x direction. We let $L = VT_p$, where, as before, T_p is the interpulse period. The correlation coefficient of powers at the two positions can be expressed as

$$\rho = \frac{E[(P_1 - \bar{P}_1)(P_2 - \bar{P}_2)]}{[\text{var}(P_1) \ \text{var}(P_2)]^{1/2}} = \frac{\sum_{i,j}' E[|A_i|^2 |A_j|^2 \cos^2 (K(x_i - x_j))]}{\sum_{i,j}' E[|A_i|^2 |A_j|^2 \cos^2 (2k(r_i - r_j))]} \tag{2.79}$$

where $P_1 = v_1 v_1^*$, $P_2 = v_2 v_2^*$, $k = 2\pi/\lambda$, and $K = 2kL/H$; r_i is the distance from the radar to the ith particle and x_i is the component of r_i along the x axis. The double summations in (2.79) are to be carried out for all i and j, for which $i \neq j$. The dependence of $|A_i|^2$ on x_i arises primarily through the gain pattern; in particular, for a rotationally symmetric Gaussian pattern with beamwidth θ_B, at nadir incidence:

$$|A_i|^2 \propto \exp(-8 \ln2 \ x_i^2/(H\theta_B)^2) \tag{2.80}$$

Carrying out the operations in (2.79) and using the above approximation for $|A_i|^2$ yields the approximation:

$$\rho(\tau) = \exp\{-[\pi\tau V\theta_B/(\lambda \ \sqrt{\ln2})]^2\} \tag{2.81}$$

where $\tau = L/V$. If we assume that successive samples are uncorrelated when $\rho(\tau) \leq 0.01$, then the decorrelation time is

$$\tau_{0.01} = 0.57 \ \lambda/V\theta_B \tag{2.82}$$

When using $\theta_B = 1.2 \ \lambda/D$, where D is the along-track antenna length, $\tau_{0.01} = 0.48 \ D/V$ which states that in order to achieve independence between successive samples, the interpulse period, T_p, should be longer than the time taken for the spacecraft to move one-half the antenna diameter. For example, for a 2 m along-track antenna dimension and $V = 7$ km/s, T_p should be greater than 137 μs, or equivalently, the PRF should be less than 7.3 kHz.

To find the standard deviation, σ, of the doppler spectrum (Hz) due to the spacecraft motion, we have for a Gaussian spectrum [24–25]:

$$\rho(\tau) = \exp\{-(2\pi\sigma\tau)^2\} \tag{2.83}$$

Equating (2.81) and (2.83) yields

$$\sigma = 0.6 \ V\theta_B/\lambda \tag{2.84}$$

which is the same as the result obtained for doppler broadening due to winds tangential to the beam [24–25].

Methods for Increasing the Number of Independent Samples

One of the most challenging aspects of spaceborne weather radar design is to provide accurate estimates of the mean return power with good resolution over a wide swath. Even if the available spacecraft power were unlimited, a reduction in the interpulse period beyond the decorrelation time has a diminishing effect on reducing the error variance of the measurement. On the other hand, if the radar pauses at a particular angle to obtain sufficient independent samples, the coverage is reduced. Some of the approaches that have been considered for increasing the effective sampling rate (number of independent samples per unit time) are listed below:

1. Pulse compression,
2. Wideband noise radar,
3. Burst mode—frequency agility,
4. Adaptive scanning,
5. Multiple-beam antennas.

In the following discussion, we briefly deal with methods 1 through 3. Scanning methods 4 and 5 are described in Section 3.3.

Pulse Compression

The advantage of pulse compression is that a monochromatic pulse of short duration τ can be simulated by a pulse of long duration and low peak power by choosing a bandwidth on the order of τ^{-1}. In contrast to altimeters and most scatterometers, the range resolution for weather radar can be relatively crude: typically 100 m or coarser because for off-nadir angles and typical beamwidths the effective horizontal and vertical resolutions are determined primarily by the beamwidth. Thus, in spaceborne weather radar, pulse compression methods are of interest not for enhancing range resolution, but for increasing the number of independent samples and avoiding the need for high peak powers.

If the desired range resolution is h (where for a monochromatic pulse $h \doteq c\tau/2$) while the range resolution of the compressed pulse is $h' \doteq c/2B'$, where B' is the bandwidth), then the effective number of samples per pulse, m, that can be attained by the use of pulse compression is h/h', or

$$m = \tau B' \tag{2.85}$$

For N pulses transmitted, the effective number of independent samples, N_e, per range interval h is given by $N_e = mN$, where we assume that the statistical properties of the scatterers in the high resolution volumes are identical.

We can find the effect on the S/N per pulse by noting that a transmitted pulse of duration τ', peak power, P_0', and bandwidth B' is equivalent to a monochromatic pulse of peak power $P'B'\tau'$ and duration $1/B'$. To compare this with the conventional case of a pulse of duration τ and peak power P_0 (where $\tau \doteq 1/B$), we have

$$S' \propto P_0'\tau'/e', \quad S \propto P_0\tau/e \tag{2.86}$$

$$N' \propto B'F', \quad N \propto F/\tau \tag{2.87}$$

where F, F' and e, e' are the respective noise figures and transmitter efficiencies for the conventional and pulse-compression cases. If the average power consumed is the same in both cases, and if the resolution for the pulse compression case is improved by a factor of m, then $\tau B' = m$ and $P_0\tau = P_0'\tau'$ so that the ratio of S/N in the conventional case to S/N in the pulse compression case is

$$\frac{(S/N)}{(S'/N')} = \frac{e'F'}{eF} m \tag{2.88}$$

For $F = F'$ and $e = e'$, this relationship implies that if the effective number of samples is increased by a factor of m, then the S/N per pulse decreases by the same factor.

If P_0' is adjusted so that the S/N is the same in both cases, then,

$$P_0'\tau' = mP_0\tau(F'e/Fe') \tag{2.89}$$

For $F = F'$ and $e = e'$, the result implies that in order to increase the sampling number by a factor of m and maintain the same S/N, an increase in the average power by a factor of m is required.

Wideband Noise Radar

The theory and operation of the noise radar for meteorology has been described by Krehbiel and Brook [26] and Doviak and Zrnić [9]. In contrast to the pulse compression methods in which the radar return is integrated coherently, the noise radar employs a phase incoherent signal (i.e., the output from a number of amplifiers incoherently summed) of bandwidth B and a radiometer-type receiver consisting of a wideband receiver, square law detector and integrator. Letting N_s be the number of particles in the scattering volume, the complex voltage envelope at the receiver input can be written in the form:

$$v = \sum_{i=1}^{N_s} a_i u(t - 2r_i/c) \exp(-i\psi_i) \tag{2.90}$$

where $u(t)$ is the transmitted baseband waveform of duration τ. Using $P = vv^*$, the variance of the power can be written [9,26] as

$$\mathrm{var}(P) = \sum_{i,j}' E(|a_i|^2 |a_j|^2) R(\tau_{ij}) \tag{2.91}$$

where the double summation extends over all i and j for which $i \neq j$, and

$$R(\tau_{ij}) = E[|u(t - 2r_i/c)|^2 |u(t - 2r_j/c)|^2] \tag{2.92}$$

where $\tau_{ij} = t_i - t_j = 2(r_i - r_j)/c$.

To simplify (2.91), we note that if the scatterers are assumed to be randomly distributed, the $|a_i|^2$ and $|a_j|^2$ are uncorrelated so that

$$E(|a_i|^2 |a_j|^2) = E(|a_i|^2) E(|a_j|^2)$$

Assuming that the distribution of particles is uniform within the volume, (2.91) can be approximated by $[(E(|a_i|^2))^2)/N_s^2] R(\tau_{ij})$. Noting that $(E(|a_i|^2))^2$ is equal to the variance of the power from a monochromatic pulse, $\mathrm{var}(P_{\mathrm{mono}})$ $(B\tau \ll 1)$, (2.91) becomes

$$(\mathrm{var}(P)/\mathrm{var}(P_{\mathrm{mono}})) = (1/N_s)^2 \sum_{ij}^{N_s}{}' R(\tau_{ij}) \tag{2.93}$$

The double sum in (2.93) can be recast in the form of an integral by the approximations:

$$(1/N_s^2) \sum_{ij}^{N_s}{}' R(\tau_{ij}) = (2/N_s^2) \sum_{i=1}^{N_s} \sum_{j=i+1}^{N_s} R(\tau_{ij}) = (2/\tau^2) \int_{t_i=0}^{\tau} \int_{t_j=0}^{t_i} R(t_i - t_j)\, dt_j\, dt_i$$

$$= (2/\tau^2) \int_0^{\tau} (\tau - x) R(x)\, dx$$

If the power spectrum associated with $|u(t)|^2$ is proportional to $\exp(-4 \ln2\, (f/B)^2)$, where B is the 6 dB bandwidth, then $R(x)$ is proportional to $\exp[-(\pi x B)^2/(4 \ln 2)]$ so that

$$[\text{var}(P)/\text{var}(P_{\text{mono}})] = 2/\tau^2 \int_0^\tau (\tau - x)R(x)\,dx \doteq \begin{cases} (\tau B)^{-1}, & B\tau \gg 1 \\ 1, & B\tau \ll 1 \end{cases} \quad (2.94)$$

The effective number of samples per pulse can be defined by

$$m = (\text{var}(P)_{B\tau \ll 1}/\text{var}(P)_{B\tau \gg 1}) \quad (2.95)$$

which, by (2.94) is approximately equal to $B\tau$.

As in the case of pulse compression, there is a direct tradeoff between S/N and the effective number of independent samples.

Burst-mode and Frequency Agility

For ground-based radar, transmission and reception of pulses are normally performed sequentially. From space, the two-way transit time, T, will generally be much greater than the interpulse period, T_p, so that the sequences of transmitted and received pulses will be displaced relative to one another. As discussed in Section 3.4, if a single antenna is used, then the PRF must be adjusted to prevent an overlap between transmission and reception. This tends to complicate the timing, especially when altitude or attitude variations are significant or when the scanning is in the cross-track direction. An alternative to this "interleaved" or sequential mode is to transmit a burst of K pulses followed by the reception of the K returns. This *burst* mode of operation was proposed in the study of a meteorological radar for the Space Shuttle for the purpose of doppler beam sharpening [27–29].

Letting T_p be the interpulse period for the pulses transmitted during the burst, τ the pulse duration, K the number of pulses per burst, and T_{pb}, the time between bursts, then the following inequalities must be satisfied to avoid overlaps between transmission and reception (see Figure 2.14(a)):

$$T_p < (T - \tau)/(K - 1) \quad (2.96)$$

$$T_{pb} > T + \tau + T_p(K - 1) > 2[\tau + T_p(K - 1)] \quad (2.97)$$

To ensure that the returns are independent, we should add the condition that T_p be larger than the decorrelation time (Equation (2.82), *et seq.*). We should also point out that these inequalities are approximate in the sense that the two-way transit time T is a function of incidence angle, altitude, attitude, *et cetera*, so that if we

a.

b.

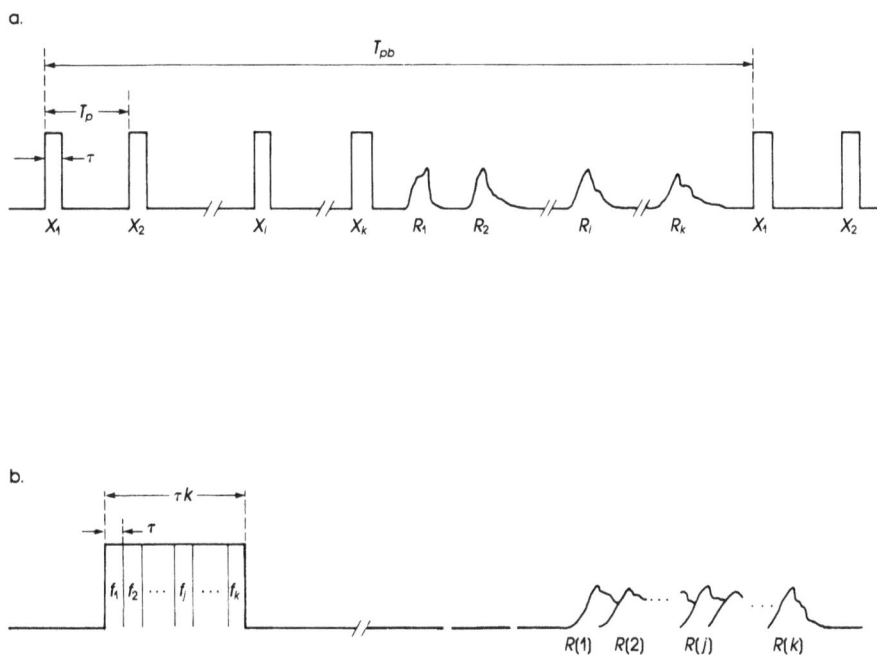

Figure 2.14 (a) Sequence of transmitted pulses $(X_j, j = 1, \ldots k)$ and the corresponding returns (R_j) for burst mode transmission; (b) Use of frequency agility and the corresponding returns $R(j), j = 1, \ldots k$ for a composite pulse of length $k\tau$.

wish to keep T_p and T_{pb} constant, T should be interpreted as the maximum transit time. Furthermore, no condition has been placed on T_p to ensure that the ith and $i+1$th returns do not overlap. This requires a more detailed discussion of the timing, which is done in Section 3.4. The use of the burst mode is a more straightforward implementation and allows greater flexibility in the choice of the PRF. On the cther hand, the sampling density generally will be smaller than that attained in the sequential mode.

To increase the sampling density, the well-known concept of frequency agility can be used [9,20,25]. A convenient way of looking at this implementation is to begin with the illustration of the burst mode timing shown in Figure 2.14(a). Keeping τ constant and letting $T_p \to 0$ leads to a composite pulse of length $k\tau$ consisting of k subpulses, each of duration τ.

To distinguish the overlapping returns from the subpulses, a modulation is introduced so that the carrier frequency of the jth subpulse is f_j [30]. A schematic of the transmitted pulse and the corresponding set of returns is shown in Figure 2.14(b); the notation $R(j)$ denotes the radar return from the jth subpulse. To ensure that the k returns are mutually independent in a statistical sense, note that

---

if the frequencies f_1 and f_2 are transmitted simultaneously and scattered from the same volume of rain, then with $\Delta f = |f_2 - f_1|$, the correlation coefficient of the powers is [20]:

$$\rho = [\sin(\pi\tau \, \Delta f)/(\pi\tau \, \Delta f)]^2 \tag{2.98}$$

so that the returns are virtually independent when $\Delta f = \tau^{-1}$ where τ is the pulse duration. Applying this result to the set of k frequencies: $f_1, f_2, \ldots f_k$ shows that the returns will be statistically independent if $|f_i - f_j| > 1/\tau$, $(i = 1, \ldots k; j = 1, \ldots k)$ for all unequal combinations of i and j.

When the above conditions are satisfied, the effective number of independent samples is increased by a factor of k over that obtained from a conventional radar (the conventional radar in this case is assumed to use a monochromatic pulse of duration $k\tau$ with the same interpulse period as that of the composite pulses). As in the previous cases, however, if the satellite power is fixed, this increase comes at the expense of a loss in the S/N per pulse. Unlike the standard pulse compression waveforms, however, this *stepped-chirp* [30] waveform uses a finite number of frequencies (more accurately, a spectrum that is highly peaked about the m frequencies) so that the surface contribution from the time sidelobes is reduced significantly. Applications of this waveform design to airborne and spaceborne meteorological radars have been discussed in the literature [30–32]. Finally, we note that the use of both frequency agility and burst-mode transmission should afford a certain amount of flexibility in satisfying the demands of timing while obtaining a sufficient number of independent samples.

Effective S/N

In general, the voltage at the input to the detector, v_T, is the sum of contributions from the rain, the surface, and the system noise [33]:

$$v_T = v_r + v_s + v_n \tag{2.99}$$

Assuming that the in-phase components of v_r, v_s, and v_n are uncorrelated zero-mean Gaussian random variables, the in-phase component of v_T is also a zero-mean Gaussian with a variance equal to the sum of the variances of the in-phase components. Identical relationships apply to the quadrature components. An estimate of power formed from N independent samples (square-law detector) is

$$\hat{P}_T = (1/N) \sum_{i=1}^{N} v_{Ti} v_{Ti}^* \tag{2.100}$$

According to the above remarks, this can be written in the form:

$$\hat{P}_T = \bar{P}_T f_N = (\bar{P}_r + \bar{P}_s + \bar{P}_n) f_N \tag{2.101}$$

where $\bar{P}_r$, $\bar{P}_s$, and $\bar{P}_n$ are the mean values of the rain, surface, and system noise powers, and where the pdf of f_N is given by (2.70).

To find the rain return power, P_r, consider the estimate:

$$\hat{P}_r = \hat{P}_T - (\bar{P}'_n + \bar{P}'_s) f_M \tag{2.102}$$

where the second term is the estimate of the surface plus noise derived from M independent samples in the absence of a rain signal.

A measure of the quality of the estimate for P_r is the effective S/N defined as the ratio of the mean of the estimate, $E(\hat{P}_r)$, to the standard deviation, $\sigma(\hat{P}_r)$. Assuming that $P'_n = P_n$ and $P_s = P'_s$, we obtain from (2.102):

$$S_e = E(\hat{P}_r)/\sigma(\hat{P}_r) = \sqrt{N}\,[(1 + 1/S/C + 1/S/N)^2$$

$$+ (N/M)(1/S/C + 1/S/N)^2]^{-1/2} \tag{2.103}$$

where $S/N = \bar{P}_r/\bar{P}_n$ and $S/C = \bar{P}_r/\bar{P}_s$ are, respectively, the signal-to-noise and signal-to-clutter ratios per pulse. In the case of large signal-to-clutter ratios, the above equation reduces to

$$S_e = S/N \sqrt{N}\,[(1 + S/N)^2 + N/M]^{-1/2} \tag{2.104}$$

Letting $M = N$, then in the limit of large S/N, the effective signal-to-noise ratio, S_e, is equal to the square root of the number of independent samples, while, for $S/N \ll 1$, S_e is equal to $S/N\,(N/2)^{1/2}$.

Trade-offs between N and S/N

As noted earlier, broadband waveforms can be used to increase the number of independent samples, N, but at the expense of the S/N per pulse. The questions that arise are what is the optimum combination of N and S/N, and how does this selection depend on the characteristics of the target. To simplify the study of these questions, we assume that for a fixed amount of power the following constraint applies:

$$N\,S/N_0 = P_a Z_0/r_0^2 = c \tag{2.105}$$

where S/N_0 is the signal-to-noise ratio per pulse at the first gate, located at r_0, which contains scatterers. Z_0 is the effective reflectivity factor at r_0 and P_a is proportional to the available spacecraft power. For a particular Z_0 and radar design, c is a constant.

In general, the S/N is a function of the range r and can be expressed as $(r \geq r_0)$:

$$S/N(r) = S/N_0[(Z(r)r_0^2/Z(r_0)r^2) \exp(-0.2 \ln 10 \int_0^r k(s) \, ds)] \tag{2.106}$$

Assuming the medium to be uniform, the equation becomes

$$S/N(r) = S/N_0(r_0/r)^2 \exp(-0.2 \ln 10 \, k(r - r_0)) \equiv S/N_0 q \tag{2.107}$$

Substituting this expression into (2.104) and using (2.105) yields

$$S_e = [1/N + 2/cq + N(1 + N/M)/(cq)^2]^{-1/2} \tag{2.108}$$

In solving for the maximum value of S_e, we set $N = M$ to avoid a cubic equation in N. Differentiating (2.108) with respect to N and setting the result equal to zero gives the value of N for which S_e is maximum:

$$N^* = (cr_0^2/\sqrt{2} \, r^2) \exp(-0.2 \ln 10 \, k(r - r_0)) \tag{2.109}$$

Substituting (2.109) into (2.105) gives

$$S/N_0 = \sqrt{2} \, (r/r_0)^2 \exp(0.2 \ln 10 \, k(r - r_0)) \tag{2.110}$$

For precipitation targets from spacecraft altitudes, $r \doteq r_0$ for all r so that if the attenuation is negligible then $N^* = c/\sqrt{2}$ and $S/N_0 = \sqrt{2}$. The corresponding S_e is then $S_e = 0.455 \sqrt{c}$.

In Figure 2.15, the quantity $10 \log S_e$, where S_e is given by (2.108), is plotted *versus* $\log N$ for three values of c, assuming that $q = 1$. That the optimum choice of N is a function of the target reflectivity suggests that S_e can be optimized by varying the relative proportions of N and S/N_0, depending on the reflectivity. In the case of weakly reflective scatterers, for example, a high sample number should be sacrificed to enhance S/N_0, whereas the opposite tradeoff should be made in cases of strong reflectivities. Such an *adaptive processing* scheme may be possible when using digital chirp or the 'stepped chirp' implementation [30]. As in the case of adaptive scanning (see Section 3.3), estimates of the target strength would be made either by a separate "look-ahead" sensor or on the basis of the first few backscattered echoes. We emphasize, however, that a radar of this type has not

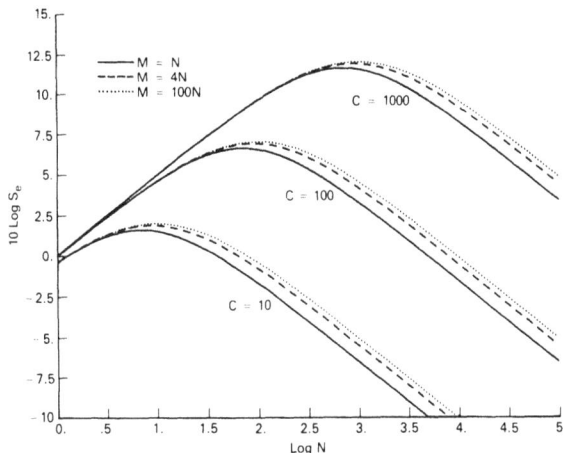

Figure 2.15 Effective signal to noise ratio, S_e (dB), *versus* the logarithm of the number of independent samples, N, with the rain signal present. M is the number of independent samples of noise only.

yet been built, and may prove impractical. A much simpler alternative would be to choose N such that S_e is optimized over the range of target reflectivities of greatest interest.

The best choice of N is further complicated when attenuation or the variation in range is significant. Figure 2.16 shows 10 log S_e curves *versus* range for the

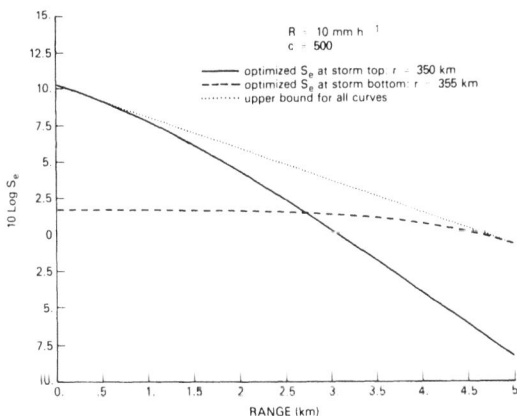

Figure 2.16 Effective signal to noise ratio, S_e (dB), *versus* depth into the storm for the cases where the number of independent samples N is chosen to optimize S_e at the storm top (solid line) and where N is chosen to optimize S_e at the surface (dashed line). The dotted line is the upper bound for all S_e curves for the case where $c = 500$ and $R = 10$ mm/h.

conditions: $c = 500$, $R = 10$ mm/h, $\lambda = 0.86$ cm and $k = 0.22\,R$ (dB/km). The solid line corresponds to a choice of N ($N = c/\sqrt{2}$) that optimizes S_e at the storm top, with $r_0 = 350$ km; the dashed line arises from selecting an N that optimizes S_e just above the surface at $r_B = 355$ km. In the latter case, the value of N is given by

$$N = [cr_0^2/\sqrt{2}\; r_B^2]\exp[-0.2\;\ln 10\; k(r_B - r_0)] \tag{2.111}$$

The upper bound for all S_e curves is indicated in Figure 2.16 by the dotted line.

The results can be interpreted in a manner similar to the previous figure in that optimizing S_e for weaker targets (caused in this case by path attenuation) requires a decrease in the number of samples. Trade-offs between N and S/N_0 are relevant to issues of maximizing penetration depth into the precipitation or cloud layer (see Section 3.7).

2.6 BISTATIC RADAR

Although the possibility of a geosynchronous weather radar has been considered by several investigators [34–35], attaining useful resolution from this altitude at microwave frequencies is well beyond present capabilities. An intriguing variation that deals with the resolution problem is the use of a bistatic geometry with a transmitter on a geosynchronous platform, and a series of ground-based receivers [36]. In one such implementation, the transmitter sequentially scans over the continental US with an *instantaneous field of view* (IFOV) of 600 km, a dwell time at each IFOV of about 2.8 minutes, and a revisit time of 90 minutes. Under the condition that the common volume is filled with rain, the S/N per pulse is calculated to be 5 dB at a 1 mm/h rainfall rate, increasing to 23 dB at a rainfall rate of 16 mm/h. Characteristics of the transmitter are: peak power of 60 kW, 5.5 μs pulse duration and a 1.3 ms interpulse period (average power of 250 W) at a wavelength of 5 cm (C-band) [36]. Although to our knowledge, this concept has not been pursued further in the literature, we present a brief outline on bistatic meteorological radars.

For the bistatic geometry shown in Figure 2.1 the scattered field, E_s, can be expressed in dyadic form as

$$\bar{E}_s = E_i \bar{f}(\hat{o},\hat{\imath})e^{-ikr}/r \tag{2.112}$$

where E_i is the magnitude of the incident field and $\bar{f}(\hat{o},\hat{\imath})$ is the dyadic scattering amplitude [2,37–39] for incident and scattering directions $\hat{\imath}$ and $\hat{o}$, respectively. If $\hat{v}_i$ and $\hat{v}_s$ are the polarization vectors associated with the incident and scattered electric fields in the plane of scattering and $\hat{h}_i$ and $\hat{h}_s$ are the corresponding vectors perpendicular to the plane of scattering where $\hat{h}_i = \hat{\imath} \times \hat{v}_i$, $\hat{h}_s = \hat{o} \times \hat{v}_s$, $\hat{h}_i = \hat{h}_s$ (Figure 2.1), then $\bar{f}(\hat{o},\hat{\imath})$ can be written in component form as

$$\bar{f}(\hat{o},\hat{\imath}) = \hat{h}_s f(\hat{h}_s,\hat{h}_i)\hat{h}_i + \hat{h}_s f(\hat{h}_s,\hat{v}_i)\hat{v}_i + \hat{v}_s f(\hat{v}_s,\hat{h}_i)\hat{h}_i + \hat{v}_s f(\hat{v}_s,\hat{v}_i)\hat{v}_i \qquad (2.113)$$

where $f(\hat{h}_s,\hat{h}_i)$ is the scattering amplitude for an incident field propagating in the direction $\hat{\imath}$ with polarization $\hat{h}_i$ scattered into the direction $\hat{o}$ with polarization $\hat{h}_s$. The other components are defined in a similar manner. The bistatic radar cross section can be written as

$$\sigma_b(\hat{o},\hat{\imath}) = 4\pi|\hat{p}_s \cdot \bar{f}(\hat{o},\hat{\imath}) \cdot p_i|^2 \qquad (2.114)$$

where $\hat{p}_s$, $\hat{p}_i$ are the (electric field) polarization states of the receiving and transmitting antennas, respectively. From Figure 2.1, $\hat{v}_s \cdot \hat{v}_i = \cos \psi$, $\hat{h}_s \cdot \hat{h}_i = 1$, $\hat{h}_s \cdot \hat{v}_i = \hat{v}_s \cdot \hat{h}_i = 0$ so that for Rayleigh scattering from spheres (wavelength much greater than the particle diameter, D):

$$\sigma_b(\hat{o},\hat{\imath}) = (\pi^5 D^6 |K|^2/\lambda^4)|\hat{p}_s \cdot (\hat{h}_s\hat{h}_i + \hat{v}_s \cos \psi \, \hat{v}_i) \cdot \hat{p}_i|^2 \qquad (2.115)$$

The radar equation can be written as

$$P_r = \frac{\lambda^2 P_0}{(4\pi)^3} \int_V \frac{\eta(\hat{o},\hat{\imath})G_t(\hat{\imath})G_r(\hat{o})e^{-[\gamma(r_1)+\gamma(r_2)]}}{r_1^2 r_2^2} |u(t - (r_1 + r_2)/c)|^2 \, dV \qquad (2.116)$$

where $\eta(\hat{o},\hat{\imath})$ is the bistatic radar reflectivity (2.6) and $\gamma(r_1)$ is the optical depth evaluated along the incident direction (and polarization), and $\gamma(r_2)$ is evaluated along the scattered direction and polarization. Assuming, for example, an incident polarization $\hat{v}_i$, with $\mathbf{f}(\hat{\imath},\hat{\imath}; \hat{v}_i)$ as the forward scattering amplitude, then, from the forward scattering theorem [2,40–41]:

$$\sigma_t = -2\lambda \, \text{Im}[\mathbf{f}(\hat{\imath},\hat{\imath}; \hat{v}_i)] \cdot \hat{v}_i = (\pi^2 D^3/\lambda) \, \text{Im}(-K) \qquad (2.117)$$

where K is given by (2.21). The rightmost equality holds in the case of Rayleigh scattering. Recall that for a continuous distribution of drop sizes, $N(D)$, the optical depth along the path from 0 to r_1 and the bistatic radar reflectivity at $r = r_1$, respectively, can be written as

$$\gamma(r_1) = 0.23 \int_{s=0}^r \int_D \sigma_t(D)N(D,s) \, dD \, ds \qquad (2.118)$$

$$\eta(\hat{o},\hat{\imath}) = \int_D \sigma_b(\hat{o},\hat{\imath})N(D) \, dD \qquad (2.119)$$

For a continuous wave transmitter, the integration volume of (2.116) is determined either by the common region of the intersecting beams or by the rain volume, whichever is smaller. In the former case, the received power can be written as [2]:

$$P_r = 6.08 \times 10^{-4} \frac{\lambda^2 P_0 G_t(\hat{\imath}) G_r(\hat{o}) \eta(\hat{o},\hat{\imath}) \theta_t \theta_r \phi_t \phi_r e^{-\gamma(r_1)-\gamma(r_2)}}{\sin \psi (r_1^2 \phi_t^2 + r_2^2 \phi_r^2)^{1/2}} \qquad (2.120)$$

where θ_t, θ_r are, respectively, the 3 dB beamwidths of the transmitting and receiving antennas in the plane of scattering and ϕ_t, ϕ_r are the corresponding antenna beamwidths in the orthogonal plane. If the illuminated volume of the receiving antenna is much smaller than that of the transmitting antenna, and if the antenna gains are expressed as a function of the beamwidths, then from (2.120) the received power is given by

$$P_r = 5.92 \times 10^{-2} \frac{\lambda^2 P_0 \eta(\hat{o},\hat{\imath}) e^{-\gamma(r_1)-\gamma(r_2)}}{r_1 \phi_t \sin \psi} \qquad (2.121)$$

which is equivalent to stating that the integration volume is proportional to $r_1 r_2^2 \phi_r \theta_r \theta_t / \sin \psi$. Note that the power is independent of the range to and the beamwidths of the receiving antenna.

More generally, for a pulsed radar, the volume integration is given by the intersection of three volumes: the first being the common volume of intersection between transmitting and receiving beams, the second the rain volume, and the third the volume between two confocal prolate ellipsoids with foci at the receiver and transmitter where the inner and outer surfaces are described by the equations $r_1 + r_2 = ct$ and $r_1 + r_2 = c(t + \tau)$ where t is the time measured from the beginning of a pulse of duration τ. Modifications to the bistatic radar equation that account for effects of partial beamfilling, attenuation effects, as well as Mie scattering are given in the literature [2,37–39].

2.7 RADAR EQUATION FOR POLARIMETRY

For backscattering ($\hat{o} = -\hat{\imath}$), the polarization conventions used for the bistatic geometry are ambiguous. For this case, vertical polarization, $\hat{v}$, is chosen to be the polarization that lies within the plane defined by the locations of the transmitter, the center of the earth, and the target of interest. Normally the radar beamwidths are sufficiently narrow so that the polarization state of the incident field is approximately constant over the illuminated volume.

In Figure 2.17, the direction of the incident field is $\hat{\imath} = \hat{k} = \hat{x} \sin \theta - \hat{z} \cos \theta$. The hydrometeor is assumed to be rotationally symmetric about the direction $\hat{s}$, where $\hat{s} = \hat{x} \sin \theta_p \cos \phi_p + \hat{y} \sin \theta_p \sin \phi_p + \hat{z} \cos \theta_p$. Following the general notation of Holt [42] but with changes to conform to a spaceborne geometry and

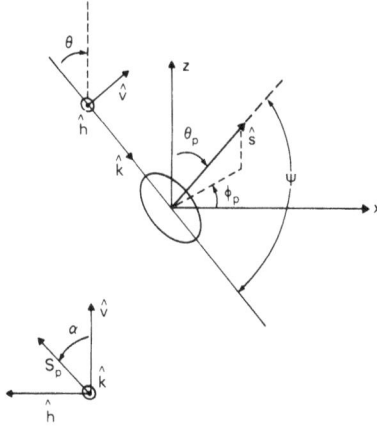

Figure 2.17 Scattering geometry for a particle with a symmetry axis along the direction s.

with an $\exp(i\omega t)$ time convention, the scattered fields that are vertically and horizontally polarized, E_{vs}, E_{hs}, can be written in terms of the scattering amplitudes and the incident fields, E_{vi} (vertical) and E_{hi} (horizontal) as

$$E_{vs} = (f_{11}E_{vi} + f_{12}E_{hi})e^{-ikr}/r \tag{2.122}$$

$$E_{hs} = (f_{21}E_{vi} + f_{22}E_{hi})e^{-ikr}/r \tag{2.123}$$

The backscattering amplitudes are given by [42]

$$f_{11} = f_{vv}(\psi) \cos^2 \alpha + f_{hh}(\psi) \sin^2 \alpha \tag{2.124}$$

$$f_{22} = f_{vv}(\psi) \sin^2 \alpha + f_{hh}(\psi) \cos^2 \alpha \tag{2.125}$$

$$f_{12} = f_{21} = [f_{vv}(\psi) - f_{hh}(\psi)] \cos \alpha \sin \alpha \tag{2.126}$$

where ψ is the angle between the incident direction, $\hat{k}$, and the symmetry axis, $\hat{s}$. The angle α is measured from the vertical polarization $\hat{v}$ to the vector s_p where s_p is the projection of $\hat{s}$ onto the polarization plane (i.e., the plane perpendicular to $\hat{k}$).

To express ψ and α as functions of the off nadir incidence angle, θ, and the orientation angles of the particle (θ_p, ϕ_p) note that

$$\hat{s} \cdot \hat{k} = \cos \psi - \sin \theta \sin \theta_p \cos \phi_p - \cos \theta \cos \theta_p \tag{2.127}$$

$$\hat{s} \cdot \hat{v} = \cos \alpha \sin \psi = \cos \theta \sin \theta_p \cos \phi_p + \sin \theta \sin \theta_p \tag{2.128}$$

$$\hat{s} \cdot \hat{h} = \sin \alpha \sin \psi = - \sin \theta_p \sin \phi_p \qquad (2.129)$$

In the literature, θ_p and α are often referred to as the canting angle and the particle canting angle, respectively.

An approximate formula for the backscattering amplitude $f_{vv}(\psi)$, valid at the lower frequencies (well below 30 GHz) or for ψ near $\pi/2$, is [42–43]

$$f_{vv}(\psi) = f(0) \cos^2 \psi + f_{vv}(\pi/2) \sin^2 \psi \qquad (2.130)$$

The equation for $f_{hh}(\psi)$ is obtained from (2.130) by replacing the subscripts v with h. In (2.130) $f(0)$ denotes the copolarized scattering amplitude when ψ is equal to 0, whereas $f_{vv}(\pi/2)$, $f_{hh}(\pi/2)$ are the copolarized returns for $\psi = \pi/2$ for vertical and horizontal polarizations, respectively. Under the Rayleigh-Gans approximation, $f(0) = f_{hh}(\pi/2)$. Similar equations apply to the forward scattering amplitudes, but over a wider range of validity. If we let $f_v(\hat{\imath},\hat{\imath}; \psi)$ denote the vertically polarized forward scattering amplitude, then [42,44]:

$$f_v(\hat{\imath},\hat{\imath}; \psi) = f(\hat{\imath},\hat{\imath}; 0) \cos^2 \psi + f_v(\hat{\imath},\hat{\imath}; \pi/2) \sin^2 \psi \qquad (2.131)$$

An identical equation, with v replaced by h, is satisfied by the horizontally polarized forward scattering amplitude, $f_h(\hat{\imath},\hat{\imath}; \psi)$.

When the wavelength is much greater than the size of the scatterer (Rayleigh-Gans approximation) $f_h(\hat{\imath},\hat{\imath}; \pi/2) = f(\hat{\imath},\hat{\imath}; 0)$, so that $f_h(\hat{\imath},\hat{\imath}; \psi) = f_h(\hat{\imath},\hat{\imath}; \pi/2)$. To evaluate these quantities for Rayleigh-Gans scattering, we begin with the equation for an ellipsoid:

$$x^2/a^2 + y^2/b^2 + z^2/c^2 = 1 \qquad (2.132)$$

For an oblate spheroid with the symmetry axis along the z direction, $c < a$, $a = b$, the volume V of the particle ($V = (4/3)\pi c a^2$) can be expressed as $\pi D_e^3/6$ where D_e is the diameter of the equivolume sphere. The backscattering amplitudes can be written as [2]:

$$f_{vv}(\pi/2) = \pi V(m^2 - 1)/[\lambda^2(1 + (m^2 - 1)L)] \qquad (2.133)$$

$$f_{hh}(\pi/2) = \pi V(m^2 - 1)/[\lambda^2(1 + 0.5(m^2 - 1)(1 - L))] \qquad (2.134)$$

where m is the complex index of refraction, and

$$L = (1 + g^{-2})(1 - g^{-1} \operatorname{Tan}^{-1} g) \qquad (2.135)$$

$$g^2 = (a/c)^2 - 1 \qquad (2.136)$$

For a prolate spheroid, $c > a$, $a = b$, $V = (4/3)\pi c a^2$, the same formulas apply, but with L given by

$$L = (e^{-2} - 1)\{-1 + 0.5\, e^{-1}\, \ln[(1 + e)/(1 - e)]\} \qquad (2.137)$$

$$e^2 = 1 - (a/c)^2 \qquad (2.138)$$

To obtain the forward scattering amplitudes, we make use of the forward scattering theorem [2] and (2.131). The extinction cross section for vertical polarization can be written:

$$\sigma_{tv} = -2\lambda\, \text{Im}[f(\hat{\imath},\hat{\imath};\, 0)\, \cos^2 \psi + f_v(\hat{\imath},\hat{\imath};\, \pi/2)\, \sin^2 \psi] \qquad (2.139)$$

which, under the Rayleigh-Gans approximation, becomes

$$\sigma_{tv} = [2\pi V\, \text{Im}(-m^2)/\lambda][|1 + 0.5(m^2 - 1)(1 - L)|^{-2}\, \cos^2 \psi$$

$$+\, |1 + (m^2 - 1)\, L|^{-2}\, \sin^2 \psi] \qquad (2.140)$$

where L is given by (2.135) for an oblate spheroid and by (2.137) for a prolate spheroid. The extinction cross section for horizontal polarization, under the Rayleigh-Gans approximation, is

$$\sigma_{th} = (2\pi V\, \text{Im}(-m^2)/\lambda)\, |1 + 0.5\, (m^2 - 1)(1 - L)|^{-2} \qquad (2.141)$$

For a sphere, $L = \tfrac{1}{3}$, and (2.140) and (2.141) reduce to the Rayleigh result: $\sigma_{tv} = \sigma_{th} = (\pi^2 D^3/\lambda)\, \text{Im}(-K)$, where D is the particle diameter and K is given by (2.21). The copolarized return power for incident vertical polarization can be written

$$P_{vv}(r) = (C_{vv} Z_{vv} |K_w|^2/r^2)\, \exp(-0.2\, \ln 10 \int_0^r k_v\, ds) \qquad (2.142)$$

where C_{vv} is the radar constant given by (2.26) where the antenna gain and beamwidths are evaluated for vertically polarized transmission and reception. Assuming a one-to-one correspondence between the equivalent drop diameter, D_e, and the eccentricity, the drop size distribution can be expressed as a function of D_e alone, $N(D_e)$, so that

$$Z_{vv} = (\lambda^4/\pi^5 |K_w|^2) \int \sigma_{vv}(\lambda, D_e;\, \psi)\, N(D_e)\, p(\theta_p, \phi_p)\, dD_e\, d\theta_p\, d\phi_p \qquad (2.143)$$

The joint pdf $p(\theta_p, \phi_p)$ for the orientation angles is assumed to be independent of the equivalent diameter of the particle. The backscattering cross section σ_{vv}

is given by $4\pi |f_{11}|^2$ where f_{11} is given by (2.124). The specific attenuation, k_v (dB per unit length) is given in terms of σ_{tv} (2.139) by:

$$k_v = 4.343 \int \sigma_{tv}(\lambda, D_e; \psi) \, N(D_e) \, p(\theta_p, \phi_p) \, dD_e \, d\theta_p \, d\phi_p \tag{2.144}$$

We note that in this approximation for the attenuation, no account is made for depolarization effects within the backscattering volume.

The cross-polarized return for incident vertical polarization can be written as

$$P_{hv}(r) = (C_{hv}Z_{hv}|K_w|^2/r^2) \exp[-0.1 \ln 10 \int_0^r (k_v + k_h) \, ds] \tag{2.145}$$

where Z_{hv} is given by (2.143) with σ_{vv} replaced by σ_{hv} where $\sigma_{hv} = 4\pi|f_{21}|^2$. The co-polarized and cross-polarized returns for incident horizontal polarization can be written in forms analogous to (2.142) and (2.145).

In the absence of attenuation, the radar reflectivity factors can be found directly from the return power measurements. The three most commonly used quantities in meteorological studies are the *differential reflectivity ratio*, ZDR, [45–46], the *linear depolarization ratio*, LDR, [42] and the *circular depolarization ratio*, CDR [47]. The ZDR and LDR are given by

$$\text{ZDR} = 10 \log (Z_{hh}/Z_{vv}) \tag{2.146}$$

$$\text{LDR} = 10 \log (Z_{vh}/Z_{hh}) \tag{2.147}$$

The CDR and related quantities are defined by Holt [42] in terms of the co-polarized and cross-polarized linear scattering amplitudes. In the simplest case, the raindrops are aligned with their symmetry axes along the vertical direction so that $\alpha = 0$ and $\psi = \pi - \theta$, so that in the absence of attenuation,

$$Z_{vv} = (\lambda^4/\pi^5|K_w|^2) \, 4\pi \int |f_{hh}(\pi/2) \cos^2 \theta$$

$$+ f_{vv}(\pi/2) \sin^2 \theta|^2 \, N(D_e) \, dD_e \tag{2.148}$$

$$Z_{hh} = (\lambda^4/\pi^5|K_w|^2) \, 4\pi \int |f_{hh}(\pi/2)|^2 \, N(D_e) \, dD_e \tag{2.149}$$

$$Z_{vh} = Z_{hv} = 0 \tag{2.150}$$

The differential reflectivity ratio, ZDR, reduces to the standard formula for $\theta = \pi/2$ and to 0 for nadir incidence. Because of the assumption of uniform alignment of the particles, the linear depolarization ratio, LDR, is $-\infty$.

2.8 DOPPLER CONSIDERATIONS

To explore doppler effects from a spaceborne weather radar, it is instructive to begin with a generalized definition of the bistatic scattering cross section from a hydrometeor of diameter D [2]:

$$\sigma_b(D,\tau) = \lim_{R\to\infty} [R^2 \langle E_s^*(t)E_s(t+\tau)\rangle] \tag{2.151}$$

where $E_s(t)$ and $E_s(t+\tau)$ are the scattered fields at the receiver at times t and $t+\tau$ and where the angular brackets denote an ensemble average. For a plane wave along the direction $\hat{\imath}$ interacting with a particle of velocity v and scattered in the direction $\hat{o}$, the cross section can be written as

$$\sigma_b(D,\tau) = \sigma_b(D)\langle\exp[-ik(\hat{\imath}-\hat{o})\cdot\mathbf{v}\,\tau]\rangle \tag{2.152}$$

where $\sigma_b(D)$, the standard bistatic cross section, is given by (2.151) with $\tau \to 0$. For nonspherical scatterers, the definition of $\sigma_b(D)$ should include averages over the orientations of the particles.

If $\mathbf{v}$ is the sum of mean and random components, $\mathbf{v} = \langle\mathbf{v}\rangle + \mathbf{v}_T$ and assuming a monostatic geometry, $\hat{\imath} = \hat{r} = -\hat{o}$, then the radar cross section becomes

$$\sigma_b(D,\tau) = \sigma_b(D)\,\exp(-2ik\hat{r}\cdot\langle\mathbf{v}\rangle\tau)\,\langle\exp(-2ik\hat{r}\cdot\mathbf{v}_T\tau)\rangle \tag{2.153}$$

For an isotropic Gaussian distribution in $\mathbf{v}_T$ with zero mean and variance σ_T^2, the last term can be evaluated to give [2]:

$$\exp[-2(2\pi\sigma_T\tau/\lambda)^2] \tag{2.154}$$

Using the above results, the radar reflectivity can be written as

$$\eta(\tau) = \int N(D)\,\sigma_b(D,\tau)\,dD$$

$$= \int_D N(D)\,\sigma_b(D)\,\exp[-2(2\pi\sigma_T\tau/\lambda)^2]\,\exp(-2ik\hat{r}\cdot\langle\mathbf{v}\rangle\tau)\,dD \tag{2.155}$$

The mean velocity of the hydrometeors relative to a spacecraft traveling along the x direction with speed v_s (neglecting the effects of the earth's rotation) is

$$\langle\mathbf{v}\rangle = \hat{x}(v_x - v_s) + \hat{y}v_y + \hat{z}[v_u - v_t(D)] \tag{2.156}$$

where v_x and v_y are the horizontal wind components, v_u is the updraft air speed, and $v_t(D)$ represents the terminal fall speed of the hydrometeors in still air.

The correlation of the baseband voltages at the receiver at times t and $t + \tau$ can be written approximately as [2]:

$$\langle V^*(t)V(t + \tau) \rangle$$

$$= \frac{\lambda^2 P_o}{(4\pi)^3} \int_V \frac{G^2(\Omega)\eta(\tau)e^{-0.2 \ln 10 \int_0^r k \, ds} \, u^*(t - 2r/c)u(t + \tau - 2r/c)}{r^4} \, dV \quad (2.157)$$

where the effects of the antenna scanning during the time τ have been neglected.

If the antenna pointing angle is (θ_0, ϕ_0) where θ_0 is the polar angle measured from nadir and ϕ_0 is the azimuthal angle measured ccw from the x axis, then by making a change of variable so that the primed coordinate system is aligned with the pulse volume, the Gaussian antenna gain can be approximated by Figure 2.18:

$$G^2(x', y') = G_0^2 \exp\{-8 \ln 2 \, [(x'/r\theta_B)^2 + (y'/r\phi_B)^2]\} \quad (2.158)$$

where r is the radar range and $G_0 = \pi^2/\theta_B\phi_B$. By extending the angular integrations over all space, the volume integral can be carried out analytically. Recognizing that the multiplicative constant in the expression for $\langle V^*(t)V(t + \tau) \rangle$ is equal to P_r/η, where P_r is the radar return power (2.19) and η is the radar reflectivity as defined above with $\tau = 0$, then,

$$\langle V^*(t)V(t + \tau) \rangle = \frac{P_r}{\eta} \int_D N(D)\sigma_b(D) \, \exp(i2\pi f_d\tau) \, \exp[-(2\pi\rho\tau)^2] \, dD \quad (2.159)$$

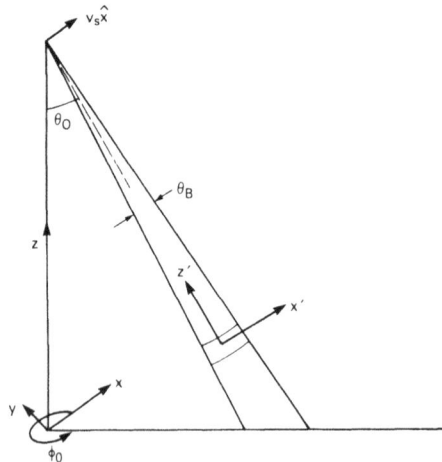

Figure 2.18 Coordinate transformation from (x,y,z) to that aligned with the radar resolution volume (x',y',z').

The mean doppler shift, f_d, is

$$f_d = (2/\lambda)[(v_s - v_x) \sin \theta_0 \cos \phi_0 - v_y \sin \theta_0 \sin \phi_0 + (v_u - v_t(D)) \cos \theta_0]$$
(2.160)

The quantity p^2, equal to one-half the variance of the spectrum, is

$$p^2 = (16 \ln2\ \sigma_T^2 + \phi_B^2 A_\phi^2 + \theta_B^2 A_\theta^2)/(8 \ln2\ \lambda^2)$$
(2.161)

with

$$A_\phi = (v_s - v_x) \sin \phi_0 + v_y \cos \phi_0$$
(2.162)

$$A_\theta = (v_s - v_x) \cos \theta_0 \cos \phi_0 - v_y \cos \theta_0 \sin \phi_0 + (v_t(D) - v_u) \sin \theta_0$$
(2.163)

The Doppler shift, f_d, can be expressed in the standard form:

$$f_d = -(2/\lambda)\hat{r} \cdot \langle \mathbf{v} \rangle$$
(2.164)

where $\hat{r}$ is the unit vector from the antenna to the scattering volume and $\langle \mathbf{v} \rangle$ is given by (2.156). Thus, a scatterer traveling away from the radar produces a negative doppler shift.

For a low earth orbit, v_s will be much greater than any component of the wind field so that, excluding nadir incidence, a good approximation for f_d is

$$f_d = (2/\lambda)v_s \cos \phi_0 \sin \theta_0$$
(2.165)

which is proportional to the cosine of the angle between $\hat{r}$ and the satellite trajectory, $\hat{x}$. We can restate this by saying that the surfaces of constant doppler shift (*isodops*) are given by the family of cones with vertices at the spacecraft and generating axes along x. The surface of zero-mean doppler is obtained in the limiting case of the plane parallel to the $y-z$ plane which passes through the spacecraft. Note that the intersection of the cones with a planar surface yields isodops which are hyperbolas [33]. The quantity p^2 is equal to the sum of components which arise from turbulence, and beam broadening in the elevation and azimuthal directions. A more general derivation of the autocorrelation reveals several other additive components to p^2, including broadening caused by antenna scanning [9], tumbling or oscillations of the drops [2], wind shear [9,25], and the Earth's rotation.

Examination of the functional dependencies of A_ϕ and A_θ shows that while only A_θ depends on the drop diameter via the terminal velocity $v_t(D)$, A_ϕ and A_θ

---

are functions not only of the winds perpendicular to the beam axis but to the antenna pointing direction. For example, in a cross-track scan, $\phi_0 = \pi/2, 3\pi/2$:

$$A_\phi = (v_s - v_x)\, \text{sgn}(\pi - \phi_0) \tag{2.166}$$

$$A_\theta = [v_t(D) - v_u]\sin\theta_0 - v_y\cos\theta_0\,\text{sgn}(\pi - \phi_0) \tag{2.167}$$

In this case, the dominant term arises from A_ϕ so that

$$p^2 = (\phi_B v_s)^2/(8\ln 2\,\lambda^2) \tag{2.168}$$

For an arbitrary scan, keeping only those terms involving v_s,

$$p^2 = (v_s)^2(\phi_B^2\sin^2\phi_0 + \theta_B^2\cos^2\theta_0\cos^2\phi_0)/(8\ln 2\,\lambda^2) \tag{2.169}$$

The corresponding decorrelation time, $\tau_{0.01}$ at the 0.01 level (i.e., $\exp - (2\pi p\tau)^2 = 0.01$) is

$$\tau_{0.01} = 0.804\,\lambda/[v_s(\phi_B^2\sin^2\phi_0 + \theta_B^2\cos^2\theta_0\cos^2\phi_0)^{1/2}] \tag{2.170}$$

At nadir incidence, the definitions of ϕ_B and θ_B are ambiguous. One way of obtaining the correct expression is by letting $\phi_0 \to \pi/2$, $\theta_0 \to 0$, and interpreting ϕ_B as the beamwidth along the direction of spacecraft motion; then,

$$\tau_{0.01} = 0.804\,\lambda/v_s\phi_B \tag{2.171}$$

This is larger than the previous expression for $\tau_{0.01}$ (see Section 2.5) by a factor of $\sqrt{2}$. The discrepancy can be resolved by noting that in the previous derivation the decorrelation time referred to power and not voltage. More generally, $\tau_{0.01}$ can be expressed as

$$\tau_{0.01} = 0.3415/p \tag{2.172}$$

where p is given by (2.161). As noted previously, additional broadening mechanisms such as turbulence, antenna rotation, and wind shear shorten the time to decorrelation (i.e., increase the spectral variance [9,25]).

Power Spectral Density

Using the usual definition of *power spectral density* (or simply *power spectrum*), $W(f)$, as the Fourier transform of the autocorrelation of V, then,

$$W(f) = \int_{-\infty}^{\infty} \langle V^*(t)V(t + \tau)\rangle \exp(-i2\pi f\tau)\,d\tau \tag{2.173}$$

Using (2.159) yields

$$W(f) = \frac{P_r}{2\sqrt{\pi}\,\eta} \int \frac{N(D)\,\sigma_b(D)}{p} \exp[-(f - f_d)^2/4p^2]\,dD \tag{2.174}$$

Noting that

$$\int_{-\infty}^{\infty} W(f)\,df = P_r \tag{2.175}$$

the normalized power spectral density, $W_n(f)$, can be written as

$$W_n(f) = W(f)/\int_{-\infty}^{\infty} W(f)\,df$$

$$= \left(\frac{1}{2\sqrt{\pi}\,\eta}\right)\int_D \frac{N(D)\,\sigma_b(D)}{p} \exp[-(f - f_d)^2/4p^2]\,dD \tag{2.176}$$

In general, both f_d and p are functions of the diameter D; however, in the case of cloud droplets, v_t is approximately zero so that (2.176) reduces to a Gaussian distribution of mean f_d and variance $2p^2$.

Estimation of Updraft Velocity and Drop Size Distribution

In all probability, the first generation of spaceborne doppler weather radars will be confined to near-nadir viewing. For this geometry, the parameter of greatest interest is the updraft velocity, v_u. Although estimation of the drop size distribution, $N(D)$, is significantly more difficult, because the two problems are closely coupled and because of the importance of the drop size distribution, we will discuss both.

To simplify the problem, we assume nadir incidence. Referring to (2.176), we have

$$\lim_{p \to 0} \exp[-(f - f_d)^2/4p^2]/p = 2\sqrt{\pi}\,\delta(f - f_d) \tag{2.177}$$

where δ is the Dirac delta function, and

$$f_d = (2/\lambda)[v_u - v_t(D)] \tag{2.178}$$

Therefore, in the limit of vanishing spectral variance, (2.176) becomes

$$W_n(f) = \frac{1}{\eta}\int_D N(D)\,\sigma_b(D)\,\delta[f - f_d(D)]\,dD \tag{2.179}$$

This relationship can also be found by setting $p = 0$ in the expression for $\langle V^*(t)\, V(t + \tau)\rangle$ and Fourier transforming.

The delta function can be transformed in the same manner as a pdf so that [2,40]

$$\delta[f - f_d(D)] = \sum_{i=0}^{N} \delta(D - D_i)/|\mathrm{d}f_d/\mathrm{d}D|_{D=D_i} \tag{2.180}$$

where D_i are the roots of the equation $f - f_d(D_i) = 0$. Assuming that $v_t(D)$ is a monotonic function of D,

$$W_n(f) = N(D_0)\, \sigma_b(D_0)/(\eta|\mathrm{d}f_d/\mathrm{d}D|_{D_0}) \tag{2.181}$$

where D_0 is the solution to

$$v_t(D_0) = v_u - \lambda f_d/2 \tag{2.182}$$

This result can be written in the more usual form by expressing the normalized spectrum as a function of the radial velocity, v, ($f = |2v/\lambda|$). Noting that

$$W_n(f) = \tilde{W}_n(v)|\mathrm{d}v/\mathrm{d}f| \tag{2.183}$$

and

$$|\mathrm{d}f_d/\mathrm{d}D| = (2/\lambda)|\mathrm{d}v_t(D)/\mathrm{d}D| \tag{2.184}$$

then,

$$\tilde{W}_n(v) = N(D_0)\sigma_b(D_0)/(\eta|\mathrm{d}v_t/\mathrm{d}D|_{D_0}) \tag{2.185}$$

In the case of Rayleigh scattering, $\sigma_b(D_0)/\eta = D_0^6/Z$; therefore,

$$\tilde{W}_n(v) = N(D_0)D_0^6/(Z|\mathrm{d}v_t/\mathrm{d}D|_{D_0}) \tag{2.186}$$

If the updraft velocity, v_u, and the functional dependence of v_t on D are known, then measurements of $\tilde{W}_n(v)$ and η provide estimates of the DSD at all diameters for which there is a corresponding measurement of $\tilde{W}_n(v)$. In particular, if $v_t(D)$ is given by a power law $v_t(D) = aD^b$ and if $W_n(f)$ is measured at frequencies $f_j = (1/2T_p)\,[2(j - 1)/(M - 1) - 1], j = 1, \ldots, M$, where T_p is the interpulse period, then in principle, $N(D)$ can be obtained for all positive values of D_j that satisfy the equation:

$$D_j = \{a^{-1}[v_u - (\lambda/4T_p)(2(j - 1))/(M - 1) - 1]\}^{1/b};\, j = 1, 2, \ldots, M \tag{2.187}$$

There are two primary sources of error in the method:

(a) The effects of spectral broadening are not included in (2.185). While this is sometimes justifiable in ground-based radar viewing stratiform rain, in spaceborne radar, the assumption is reasonable only for extremely narrow beamwidths or for a geosynchronous satellite.

(b) Small errors in the updraft velocity can lead to large errors in the rain rates or liquid water content derived from the estimated $N(D)$ [48–49].

To deal with the latter problem, Hauser and Amayenc [50–51] have proposed a method whereby the two parameters of an exponential model for $N(D)$ and the updraft velocity are derived simultaneously by fitting to the measured spectrum. Using $N(D) = N_0 \exp(-\Lambda D)$, Λ and v_u are found to be independent of biases in the reflectivity factor Z as these parameters depend only on the shape of the spectrum and its position along the velocity axis. On the other hand, all three parameters are functions of the statistical fluctuations in Z caused by finite sampling. An error analysis shows that a 14% error in Z leads to errors in N_0, Λ, and v_u of 30%, 5%, and 0.7 m/s, respectively.

More recently, a somewhat similar method has been proposed [52] for a ground-based doppler radar at 94 GHz. In contrast to most doppler radars, the assumption that $\sigma_b/\eta = D^6/Z$ no longer applies; in fact, the first minimum of $\sigma_b(D)$ at this frequency, which occurs at $D = 1.67$ mm, provides a precise measurement of the vertical velocity in this drop size interval. By using the terminal fall velocity from the data of Gunn and Kinzer [53], the updraft velocity can be deduced. Procedures are also described which account for spectral broadening and inferring $N(D)$. Although new, judging by the excellent results that have been obtained to date, the technique appears to be useful and potentially applicable to spaceborne doppler radar.

Estimation of the updraft velocity is simpler because the estimate depends only on the first moment of the spectra. Moreover, in the case of a symmetric spectrum, the mean doppler velocity is relatively insensitive to spectral broadening. To outline how v_u can be estimated, we note that if $\bar{v}$ is the mean velocity as derived from the measured spectrum, then an estimate of the updraft velocity can be found from [48]:

$$v_u = v_0 + \bar{v} \tag{2.188}$$

where v_0 is the mean doppler velocity that would be measured in the absence of updrafts. The sign convention used here is that positive $\bar{v}$ is associated with velocities toward the radar ($+ z$ axis). On the other hand, by convention, positive v_0 is associated with a velocity along the $- z$ axis or toward a ground-based radar. To estimate v_0, let $N(D)$ be given by a gamma distribution (see Section 4.3) and $v_t(D)$ by a power law (see Section 4.4), i.e.,

$$N(D) = N_0 D^m \exp(-\Lambda D) \tag{2.189}$$

$$v_t(D) = aD^b \tag{2.190}$$

By definition,

$$v_0 = \left| \frac{\lambda}{2} \int f \, W_n(f) \right|_{v_u=0} df = \left| \int v \, W_n(v) \right|_{v_u=0} dv \tag{2.191}$$

Using (2.186) and making a change of variable to D_0 yields

$$v_0 = aN_0\Gamma(7 + b + m)/Z\Lambda^{(7+b+m)} \tag{2.192}$$

where Γ is the complete gamma function. Moreover,

$$Z = \int D^6 N(D) \, dD = N_0\Gamma(7 + m)/\Lambda^{(7+m)} \tag{2.193}$$

Eliminating Λ between these equations gives

$$v_0 = aK(Z/N_0)^{(b/(7+m))} \tag{2.194}$$

where K is a function only of b and m. Noting that Z and $\bar{v}$ are measured quantities, v_u can be obtained from (2.188) and (2.194) if the parameters a, b, N_0 and m are assumed. We note that the value of v_0 is sometimes chosen *a priori* (e.g., $v_0 = 1$ m/s for snow and $v_0 = 0$ for cloud). Another approach has been to estimate the updraft by recognizing that if the radar is sufficiently sensitive, the terminal velocities of the minimum detectable raindrops are small so that, apart from a small correction factor, the measured velocity of the small raindrops is approximately equal to the updraft [54].

One of the most common $v_0 - Z$ relationships for rain is obtained by taking $N_0 = 0.08$ cm^{-4}, $m = 0$ and $a = 1430$ cm$^{-1/2}$ s^{-1}, and $b = 0.5$ in (2.194) which yields [55–56]:

$$v_0 = 3.84 \, Z^{0.0714} \tag{2.195}$$

where v_0 is in m/s and Z in mm^6/m^3. A discussion of $v_0 - Z$ relationships for rain and hail can be found in Rogers [48] and Ulbrich [57]. According to Rogers, for ground-based radars observing widespread midlatitude moderate rain, where the above $v_0 - Z$ relationship is representative, the estimates of v_0 are normally accurate to within ± 1 m/s.

Nyquist Criterion

In the foregoing discussion we have assumed that the autocorrelation function is available for all values of τ. In fact, the function is measured only at integral multiples of the interpulse period, T_p, so that $\tau = mT_p$, $m = 1, \ldots, M$. From the set of M samples, the spectrum can be determined at frequencies f_j:

$$f_j = (1/2T_p)[2(j - 1)/(M - 1) - 1] \tag{2.196}$$

To avoid aliasing, the quantity $1/2T_p$ must be greater than the largest frequency in the spectrum. For Gaussian spectra, this is not possible because the spectral width is unbounded. If we assume momentarily that f_d is zero or is known, then as a practical condition we require that the standard deviation, $\sqrt{2}\,p$ (where p^2 is given by (2.161), is less than $1/2T_p$. Using $f_{PRF} = 1/T_p$,

$$f_{PRF} \geq 2\sqrt{2}\,p \tag{2.197}$$

To avoid range ambiguities, the PRF must also satisfy the condition:

$$f_{PRF} \leq c \cos \theta_0/2H_s \tag{2.198}$$

where H_s is the maximum height of detectable hydrometeors. Eliminating f_{PRF} in these inequalities, and keeping only the terms of p which are functions of the satellite speed v_s yields

$$(\phi_B^2 \sin^2 \phi_0 + \theta_B^2 \cos^2 \theta_0 \cos^2 \phi_0)^{1/2} \leq 0.416\, c\, \lambda \cos \theta_0/v_s H_s \tag{2.199}$$

To get an idea of the antenna diameter, D_a, needed to satisfy this condition, let H_s = 18 km and v_s = 7 km s^{-1}. Using a symmetric antenna pattern with $\theta_B = \phi_B$ = $1.3\lambda/D_a$ then at nadir incidence D_a must be greater than 1.3 meters. It is worth noting that D_a is independent of wavelength.

The above inequality is only an approximation. By staggering the interpulse period, for example, (2.197) can be relaxed. On the other hand, for most radar designs, the PRF must be reduced to less than the theoretical maximum given by (2.198). Moreover, f_d is not completely known, and in fact contains the principal quantities of interest. Letting σ_s^2 be the sum of squares of the rms values of the three wind components, and denoting the errors in the antenna pointing angle by $\Delta\theta_0$ and $\Delta\phi_0$, a more accurate expression for the range-velocity ambiguity condition is obtained by replacing p in the above equations by

$$\{p^2 + (2/\lambda)^2([v_s(\Delta\theta_0 \cos \theta_0 \cos \phi_0 - \Delta\phi_0 \sin \phi_0 \sin \theta_0)]^2 + \sigma_s^2)\}^{1/2} \tag{2.200}$$

To look at the question of sampling in a more quantitative manner, the estimate of the autocorrelation for the *pulse-pair processor* in the case of M uniformly spaced returns is [9]:

$$\hat{R}(T_p) = (1/M) \sum_{m=1}^{M-1} V^*(m)V(m+1) \tag{2.201}$$

where $V(m)$ represents the baseband voltage measured at a multiple m of the interpulse period, T_p. We can see from the equation that the autocorrelation is estimated on the basis of a sum of autocorrelations from adjacent samples. The mean velocity estimate can be obtained from the equation [9]:

$$v = -(\lambda/4\pi T_p)\, \arg(\hat{R}(T_p)) \tag{2.202}$$

where arg is the argument of (2.201) and where the minus sign is used in accordance with the convention that the velocity in the outgoing radial direction is taken to be positive (negative doppler shift).

For large S/N, M equally spaced samples, and a Gaussian spectrum, the variance of the above estimate can be written as [9]:

$$\text{var}(v) = \lambda^2[1 - \rho^2(T_p)]/[CMT_p^3\rho^2(T_p)p] \tag{2.203}$$

where

$$C = 128\pi^{2.5}/\sqrt{2} = 1.58 \times 10^3 \tag{2.204}$$

$$\rho(T_p) = \exp[-(2\pi pT_p)^2] \tag{2.205}$$

Equation (2.203) is valid under the conditions that [9]:

$$2\sqrt{2}\pi MT_p p \gg 1 \tag{2.206}$$

$$\rho^2(T_p)M \gg 1 \tag{2.207}$$

If the Nyquist criterion is satisfied in a "strong" sense such that $2\pi pT_p \ll 1$, then $1 - \rho^2(T_p) = 8(\pi pT_p)^2$ and the variance of v reduces to

$$\text{var}(v) = \lambda^2 p/(8\sqrt{2\pi}\, MT_p) = \lambda v_s \phi_B/(32\sqrt{\pi \ln 2}\, MT_p) \tag{2.208}$$

where p is given by (2.161). The rightmost equality is obtained by keeping only those terms of p that are functions of v_s and assuming nadir incidence with an along-track beamwidth ϕ_R. We note that under similar conditions (high S/N, narrow spectra and large M), this expression for the variance is also valid for the *fast Fourier transform* (FFT) processor [9].

Graphs of the standard deviation of the velocity estimate based on (2.203) (solid curves) and (2.208) (dashed curves) *versus* wavelength are shown in Figure 2.19 for three values of the along-track beamwidth where $T_p = 0.2$ ms and $M = 250$. The curves have been plotted only for those values of the parameters for which (2.206) and (2.207) are valid; to satisfy these inequalities, "much greater than" has been interpreted to be a factor of 10 or larger.

When valid, (2.208) shows that the variance of the estimate decreases if the dwell time, MT_p, increases or if the beamwidth or wavelength decreases. If the beamwidth is fixed, the variance is inversely proportional to the frequency, whereas if the physical antenna size is kept constant, the variance is inversely proportional to the square of the frequency. In either case, a reduction in the variance of the estimate is attained by use of a higher frequency. We emphasize, however, that these results are valid only when $2\pi p T_p \ll 1$, a condition which for most spaceborne radar designs is virtually independent of the frequency. This explains why the improvement predicted by (2.208) is not as large as that given by the more general expression (2.203). Finally, we note that as the spectral width broadens (i.e., (2.203) is no longer satisfied) the FFT processor provides estimates of the velocity which have a smaller variance than those of the autocorrelation receiver [9]. In the selection of the receiver for a spaceborne doppler radar, this advantage of the FFT processor must be weighed against the increases in complexity and computation time.

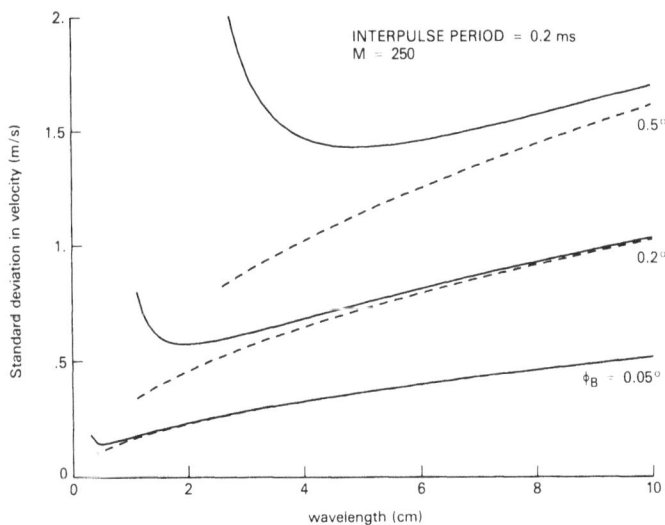

Figure 2.19 Standard deviation of the mean velocity estimate *versus* wavelength as predicted by (2.203) (solid line) and (2.208) (dashed line) for three values of the along-track beamwidth, ϕ_B.

Conical Geometry

(a) *VAD Method:* A spaceborne radar that conically scans about nadir provides a measurement geometry which in some sense is similar to that used for the *velocity-azimuth display* VAD method [9,58–59]. For ground-based radars, the VAD has been shown to be useful in obtaining mean horizontal winds over the measurement area. Experienced interpreters of the signatures have been able to deduce a variety of kinematic features of the storm system.

Error analyses have shown that the best accuracies are obtained in cases of low elevation angles over small areas and under uniform rain conditions. As these requirements are impossible to satisfy from space, the use of the VAD method is highly questionable. However, the success of this and related methods for ground-based radars suggests a suspension of judgement until a thorough analysis for spaceborne geometries has been done. We will not attempt that here, but offer an outline of the method.

In a conical scan, the polar angle, θ_0, is fixed so that the measurements of the radial velocity as a function of the azimuthal angle at a constant radar range can be written as

$$v_r(\phi) = v \cdot \hat{r} = (v_x - v_s) \cos \phi \sin \theta_0 + v_y \sin \phi \sin \theta_0 + (v_t - v_u) \cos \theta_0$$
$$(2.209)$$

where the diameter of the circle is $2(H - z) \tan \theta_0$, where H is the satellite altitude and z is the height of the plane above the surface. If the horizontal wind, v_h, is uniform over the circle and makes an angle α with respect to the x axis, then (2.209) can be written as

$$v_r(\phi) = v_h \sin \theta_0 \cos(\alpha - \phi) - v_s \cos \phi \sin \theta_0 + (v_t - v_u) \cos \theta_0 \quad (2.210)$$

If v_h and $(v_t - v_u)$ are constant over the area, then the right-hand side of (2.209) is the sum of even, odd, and constant terms. Thus, an estimate of $(v_t - v_u)$ can be found from an integration of $v_r(\phi)$ over the circumference of the circle while $(v_x - v_s)$ and v_y are obtained, respectively, by forming the sum and difference terms: $v_r(\phi) + v_r(-\phi)$ and $v_r(\phi) - v_r(-\phi)$. Once v_t and v_s are estimated, v_h, α and v_u can be determined.

A more general approach to the problem begins by using a Taylor expansion of v_r about the center of the circle. The analysis shows that the first harmonic component of $v_r(\phi)$ is equal to v_h and α averaged over the circle while the second harmonic provides a measure of the wind deformation [9]. The integral of $v_r(\phi)$ over the circumference of the circle is related to the mean divergence of the horizontal wind field [5].

(b) *Stereoscopic Viewing:* A conically scanning antenna that provides continuous coverage of the scene will view a particular volume of rain first in the forward portion of the scan ($|\phi| < \pi/2$), and a time later in the backward portion ($|\phi| > \pi/2$). To visualize the geometry, consider two identical right circular cones with half-vertex angles θ_0 and with vertices coincident with the spacecraft at positions: $(-D,0,H)$ and $(D,0,H)$ (Figure 2.20). The intersection of the surfaces in the $x = 0$ plane is the hyperbola:

$$y^2 + D^2 = \tan^2 \theta_0 (H - z)^2 \tag{2.211}$$

where z is measured from the surface and y is the distance measured from the subsatellite track.

At $D = 0$, where the surfaces of the two cones are identical, the intersection of the cone with the $x = 0$ plane are intersecting lines with interior angle $2 \theta_0$. As the separation D increases, the family of hyperbolas generated by the intersection of the cones in the $x = 0$ plane fill the area beneath these lines. With D fixed, the intersection points of the hyperbola (in the $x = 0$ plane) with spheres of radius r centered at the two spacecraft locations are given by (Figure 2.20):

$$(x,y,z) = (0, \pm r \sin \theta_0 \sin \phi, H - r \cos \theta_0) \tag{2.212}$$

where D and ϕ are related by

$$D = r \sin \theta_0 \cos \phi \tag{2.213}$$

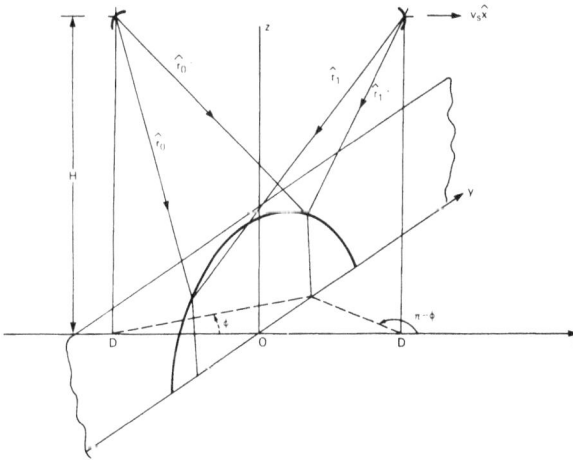

Figure 2.20 Common points of observation in a conical scanning geometry.

The unit vectors from the two spacecraft locations at $(-D,0,H)$ and $(D,0,H)$ to the points of intersection are, respectively,

$$\hat{r}_0(\pm) = (\sin \theta_0 \cos \phi, \ \pm \sin \theta_0 \sin \phi, \ -\cos \theta_0) \tag{2.214}$$

$$\hat{r}_1(\pm) = (-\sin \theta_0 \cos \phi, \ \pm \sin \theta_0 \sin \phi, \ -\cos \theta_0) \tag{2.215}$$

so that at each intersection point, the measurements are obtained for the velocities, v_0 and v_1 along the directions $\hat{r}_0$ and $\hat{r}_1$, respectively.

To resolve the velocities into rectangular components, local coordinates can be defined such that $\hat{x}_c$, $\hat{z}_c$ are in the plane defined by $\hat{r}_0$ and $\hat{r}_1$ and $\hat{y}_c$ is orthogonal to it. One such choice is

$$[\hat{x}_c, \hat{y}_c, \hat{z}_c] = [\hat{x}, (\hat{r}_1 \times \hat{r}_0)/|\hat{r}_1 \times \hat{r}_0|, \hat{x} \times \hat{y}_c] \tag{2.216}$$

The components of the velocities v_{xc}, v_{zc} along the directions $\hat{x}_c$ and $\hat{z}_c$, respectively, are given in terms of the measurable velocities v_0 and v_1 by

$$v_{zc} = (v_0 + v_1)/[2 \cos(\psi/2)] \tag{2.217}$$

$$v_x - v_s = v_{xc} = (v_0 - v_1)/[2 \sin(\psi/2)] \tag{2.218}$$

where ψ is the angle between $\hat{r}_0$ and $\hat{r}_1$ with

$$\cos(\psi/2) = (\sin^2 \theta_0 \sin^2 \phi + \cos^2 \theta_0)^{1/2} \tag{2.219}$$

Use of the continuity equation for mass conservation along with boundary conditions at the surface and storm top can help infer the remaining wind component.

This discussion has focused on dual-doppler measurements by the use of a conical scan. Recently, a modified cross-track scan has been proposed for this type of measurement which has the advantage of a more regular sampling geometry for the analysis of dual-doppler returns [60] (see Section 3.7).

REFERENCES

[1]   Silver, S., ed., 1949: Microwave Antenna Theory and Design. MIT Radiation Lab. Series, **12**, McGraw-Hill, New York, 623 pp.
[2]   Ishimaru, A., 1978: *Wave Propagation and Scattering in Random Media*. Academic Press, New York. Vol. 1, 250 pp.
[3]   Probert-Jones, J.R., 1962: The radar equation in meteorology. *Quart. J. Royal Meteor. Soc.*, **88**, 485–495.
[4]   Stephens, J.J., 1966: A note on the idealized radar range equation. *J. Appl. Meteor.*, **5**, 893–895.
[5]   Battan, L.J., 1973: *Radar Observations of the Atmosphere*, Univ. of Chicago Press, Chicago, 324 pp.

[6]   Nakamura, K., 1989: A comparison of rain retrievals by backscattering measurement and attenuation measurement. *Proc. 24th Conf. Radar Meteor.*, March 27–31, Amer. Meteor. Soc., Boston, 689–692.

[7]   Amayenc, P., M. Marzoug, and J. Testud, 1989: Nonuniform beamfilling effects in measurements of rainfall rate from a spaceborne radar. *Proc. 24th Conf. Radar Meteor.*, March 27–31, Amer. Meteor. Soc., Boston, 569–572.

[8]   Nathanson, F.E., and P.L. Smith, 1972: A modified coefficient for the weather radar equation. Preprints *15th Conf. on Radar Meteor.*, Amer. Meteor. Soc., Boston, pp. 228–230.

[9]   Doviak, R.J. and D.S. Zrnić, 1984: *Doppler Radar and Weather Observations.* Academic Press, New York, 458 pp.

[10]   Kozu, T., 1989: Consideration of vertical resolution for near nadir-looking spaceborne radar. *IEEE Trans. Geosci. and Remote Sens.*, **GE-27**, 354–357.

[11]   Ulaby F.T., 1980: Vegetation clutter model. *IEEE Trans. Ant. and Propag.*, **AP-28**, 538–545.

[12]   Ulaby, F.T., R.K. Moore, and A.K. Fung, 1981: *Microwave Remote Sensing: Active and Passive.* Artech House, Norwood, MA, Vol. 1, 456 pp.

[13]   Jones, W.L., L.C. Schroeder, and J.L. Mitchell, 1977: Aircraft measurements of the microwave scattering signature of the ocean. *IEEE Trans. Ant. and Propag.*, **AP-25**, 52–61.

[14]   Manabe, T., and T. Ihara, 1988: A feasibility study of the rain radar for the tropical rainfall measuring mission: 5. Effects of surface clutter on rain measurements from satellite. *J. Comm. Research Lab.*, **35**, 163–181.

[15]   Grant, C.R., and B.S. Yaplee, 1957: Back scattering from water and land at centimeter and millimeter wavelengths. *Proc. IRE*, **45**, 972–982.

[16]   Atlas, D. and T.J. Matejka, 1985: Airborne Doppler radar velocity measurements of precipitation seen in ocean surface reflection. *J. Geophys. Res.*, **90**, 5820–5828.

[17]   Meneghini, R., and D. Atlas, 1986: Simultaneous ocean cross section and rainfall measurements from space with a nadir-looking radar. *J. Atmos. and Oceanic Technol.*, **3**, 400–413.

[18]   Meneghini, R., and K. Nakamura, 1988: Some characteristics of the mirror-image return in rain. *Proc. Intl. Symp. on Tropical Precip. Meas.*, Tokyo, October 28–30, 235–242.

[19]   Blake, L.V., 1970: Prediction of Radar Range. Ch. 2 in *Radar Handbook*, M. I. Skolnik, ed., McGraw-Hill, New York.

[20]   Marshall, J.S., and W. Hitschfeld, 1953: Interpretation of the fluctuating echo from randomly distributed scatterers. Pt. I, *Canadian J. Phys.*, **31**, 962–994.

[21]   Stogyrn, A., 1975: Error Analysis of the Goldhirsh-Katz method of rainfall determination by the use of a two frequency radar. Report No. 1833TR-4, Aerojet Electrosystems Co., Asusa, CA.

[22]   Davenport, W.B., Jr., and W.L. Root, 1958: *Random Signals and Noise.* McGraw-Hill, New York.

[23]   Walker, G.B., P.S. Ray, D. Zrnić, and R. Doviak, 1980: Time, angle, and range averaging of radar echos from distributed targets. *J. Appl. Meteor.*, **19**, 315–323.

[24]   Hitschfeld, W., and A.S. Dennis, 1956: Turbulence in Snow Generating Cells. Sci. Rept. MW-23, McGill University, Montreal, 31 pp.

[25]   Atlas, D., 1964: Advances in radar meteorology. *Adv. Geophys.*, **10**, 317–478.

[26]   Krehbiel, P.R., and M. Brook, 1979: A broad-band noise technique for fast scanning radar observations of clouds and clutter targets. *IEEE Trans. Geosci. Electron.*, **17**, 196–204.

[27]   Eckerman, J., 1975: Meteorological radar facility for the space shuttle. *IEEE National Telecomm. Conf.*, New Orleans, IEEE publ. 75 CH1015 CSCB, 37-6–37-17.

[28]   Bucknam, J., R.P. Dooley, A. Fredriksen, and F.E. Nathanson, 1975: Synthetic Azimuth Processing Mode—Concept Analysis, Technical Note 3, Rev. 1, Technology Service Co., NASA/GSFC Contract NAS5-20058, March.

[29]   Bucknam, J., R.P. Dooley, and F.E. Nathanson, 1975: Shuttle Meteorological Radar Study, Final Report, Technology Service Co., NASA/GSFC Contract NAS5-20058, April.

[30]   Frush, C., 1989: Stepped-chirp transmit waveforms for improved fast scanning weather radars. *Proc. 24th Conf. Radar Meteor.*, March 27–31, Amer. Meteor. Soc., Boston, 439–442.

[31]   Im, K., F.K. Li, W.J. Wilson, and D. Rosing, 1987: Conceptual design of a spaceborne radar for global rain mapping. *Proc. IGARSS*, Ann Arbor, MI.

[32]   Li, F., K. Im, W.J. Wilson, and C. Elachi, 1988: On the design issues for a spaceborne rain mapping radar. *Tropical Rainfall Measurements*, J.S. Theon and N. Fugono, eds., A Deepak Publishing, Hampton, VA, 235–242.

[33]   Ulaby, F.T., R.K. Moore, and A.K. Fung, 1982: *Microwave Remote Sensing: Active and Passive*. Artech House, Norwood, MA, Vol. 2, 608 pp.

[34]   Skolnik, M.I., 1974: The Application of Satellite Radar for the Detection of Precipitation. NRL Rept. 2896, October, 100 pp.

[35]   Matthews, R.E., ed., 1975: *Active Microwave Workshop*. NASA SP-376, 502 pp.

[36]   Nathanson, F.E., 1981: Bistatic radar meteorological satellite. Precipitation Measurements from Space: Workshop Rept., Atlas, D., and O.W. Thiele, eds., NASA/GSFC, Greenbelt, MD, pp. D-341–D-350.

[37]   Crane, R.K., 1972: Virginia precipitation scatter experiment—data analysis. NASA/GSFC X-750-73-55; Revised, October 1973.

[38]   Crane, R.K., 1974: Bistatic scatter from rain. *IEEE Trans. on Antennas and Propagation*, **AP-22**, 312–320.

[39]   Awaka, J., and T. Oguchi, 1982: Bistatic radar reflectivities of Prupapacher-Pitter form raindrops at 34.8 GHz. *Radio Sci.*, **17**, 269–278.

[40]   Born, M., and E. Wolf, 1980: *Principles of Optics*, sixth ed., Pergamon Press, Oxford, 808 pp.

[41]   Van de Hulst, H.C., 1957: *Light Scattering by Small Particles*. John Wiley and Sons, New York, 470 pp.

[42]   Holt, A.R., 1984: Some factors affecting the remote sensing of rain by polarization diversity radar in the 3- to 35-GHz frequency range. *Radio Sci.*, **19**, 1399–1412.

[43]   Holt, A.R., and J.W. Shepherd, 1979: Electromagnetic scattering by dielectric spheroids in the forward and backward directions. *J. Phys. A.*, **12**, 159–166.

[44]   Uzunoglu, N.K., B.G. Evans, and A.R. Holt, 1977: Scattering of electromagnetic radiation by precipitation particles and propagation characteristics of terrestrial and space communications systems. *Proc. Inst. Electr. Eng.*, **124**, 417–424.

[45]   Seliga, T.A., and V.N. Bringi, 1976: Potential use of radar differential reflectivity measurements at orthogonal polarizations for measuring precipitation. *J. Appl. Meteor.*, **15**, 69–76.

[46]   Seliga, T.A., and V.N. Bringi, 1978: Differential reflectivity and differential phase shift: Applications in radar meteorology., *Radio Sci.*, **13**, 271–275.

[47]   McCormick, G.C., and A. Hendry, 1985: Principles for the radar determination of the polarization properties of precipitation. *Radio Sci.*, **10**, 421–434.

[48]   Rogers, R.R., 1982: A review of multiparameter radar observations of precipitation. *Radio Sci.*, **19**, 23–36.

[49]   Atlas, D., R.C. Srivastava, and R.S. Sekon, 1973: Doppler radar characteristics at vertical incidence. *Rev. Geophys. Space Phys.*, **2**, 1–35.

[50]   Hauser, D., and P. Amayenc, 1981: A new method for deducing hydrometeor-size distributions and vertical air motions from Doppler radar measurements at vertical incidence. *J. Appl. Meteor.*, **20**, 547–555.

[51]   Hauser, D., and P. Amayenc, 1981: Exponential size distributions of raindrops and vertical air motions deduced from zenith-pointing doppler radars in a frontal precipitation. *Proc. 20th Conf. Radar Meteor.*, Amer. Meteor. Soc., Boston, 91–98.

[52]   Lhermitte, R., 1989: Mie scattering observations by a 94 GHz doppler radar at vertical incidence. *Proc. 24th Conf. Radar Meteor.*, March 27–31, Amer. Meteor. Soc., Boston, 1–4.

[53] Gunn, K.L.S., and G.D. Kinzer, 1949: The terminal velocity of fall for water droplets in stagnant air. *J. Meteor.*, **6**, 243–248.

[54] Battan, L.J., 1964: Some observations of vertical velocities and precipitation sizes in a thunderstorm. *J. Appl. Meteor.*, **3**, 415–420.

[55] Marshall, J.S., and W.M.K. Palmer, 1948: The distribution of raindrops with size. *J. Meteor.*, **5**, 165–166.

[56] Spilhaus, A.F., 1948: Drop size intensity and radar echo of rain. *J. Meteor.*, **5**, 161–164.

[57] Ulbrich, C.W., 1977: Doppler radar relationships for hail at vertical incidence. *J. Appl. Meteor.*, **16**, 1349–1359.

[58] Lhermitte, R., and D. Atlas, 1961: Precipitation motion by pulse doppler radar. *Proc. Ninth Weather Radar Conf.*, Amer. Meteor. Soc., Boston, 218–223.

[59] Doviak, R.J., and R.G. Strauch, 1980: Pt III. Single radar data acquisition. The multiple doppler radar workshop. *Bull. Amer. Meteor. Soc.*, **10**, 1178–1183.

[60] Lhermitte, R., 1989: Satellite-borne millimeter wave doppler radar. URSI Commission F, Open Symposium, September 11–15, La Londe-Les-Maures, France.

Chapter 3
Design Considerations

3.1 OVERVIEW

The quality and quantity of precipitation data that are needed for most applications far exceed the capabilities of any single instrument. Consequently, the orbiting weather radar should be viewed as one of a set of instruments. Nevertheless, in considering the radar design, it is instructive to investigate a wide range of radar types that might satisfy demands such as high resolution, wide coverage or high sampling rates. In this chapter, we discuss some of the design issues in spaceborne weather radar, including orbital considerations, frequency selection, and scanning options. This leads to the topic of spacecraft hardware and tradeoffs in the choice of antenna and transmitter. In Section 3.6, we summarize some of the work done in the use of *synthetic aperture radar* (SAR) for meteorological sensing. We conclude the chapter with a review of some recent spaceborne weather radars that have been proposed.

To appreciate the choices in design, it is helpful to be aware of some of the constraints. For a "monochromatic" pulsed radar, the PRF is limited by the condition that only one pulse be present in the scattering medium at a given time. Other constraints apply if, for example, the same antenna is used for transmission and reception or if statistical independence is desired from pulse to pulse. The PRF is also affected by the possible presence of the mirror-image return, altitude and attitude variations of the spacecraft, and the scan mode. Together with dwell time, the PRF determines the number of samples N within the resolution cell. Specifying some minimum N is difficult, as the dynamic range and accuracy of the meteorological estimate also depend upon power, frequency, antenna gain and particular characteristic of the precipitation that is to be estimated. Nevertheless, for most quantitative estimates of rain parameters, the minimum N is typically on the order of 30, which gives a standard deviation in the return power estimate of about 1 dB. For the application of dual-wavelength, polarimetry, or doppler methods, however, larger values of N are generally required. Several techniques are available to increase N without the loss of coverage. These include pulse compression methods, the wideband noise method and frequency agility (see Section 2.5). In all these methods, however, the increase in the effective N is at the expense of decreasing the S/N per pulse.

Attaining sufficient sampling at each resolution cell still leaves open the question of coverage. Methods for achieving a wide swath have received a great deal of attention in the design of spaceborne weather radars. Among the systems that have been studied are the fan-beam (side-looking) radar, pencil-beam antennas used either in a conical or cross-track scan, multiple-beam antennas, and synthetic aperture. Hybrids have also been analyzed, including azimuthal beam-sharpening with multiple receiving beams, and multiple noncontiguous beams with subswath scans. Like broadband waveform designs, the use of multiple beams becomes attractive if power can be traded for wider coverage or increased sampling. The concept of adaptive scanning starts with the premise that the available time and radar power should be allotted on the basis of the importance of the target. A number of implementations of this idea have been proposed, ranging from a relatively simple power-saving method to sophisticated dual-level scans that use a rapid search for detection, and possibly classification of rain intensities, followed by slower scan throughout the regions of greatest interest.

Although radar design deals directly with the issues of sampling and scanning, to the meteorologist and climatologist, a question of greater importance is the spatial and temporal sampling over the storm system. This is a function of not only the radar parameters, but also the altitude and orbital inclination. For a 30° orbital inclination and a satellite altitude of 350 km, the distance along adjacent orbits near the equator is about 2500 km. Thus, a swath of this dimension would be needed to eliminate gaps in the coverage. In the case of a polar orbiter, the full cross-track scan angle, 2θ, needed to provide continuous coverage (without gaps) at the equator is [1]

$$2\theta = 2\,\text{Tan}^{-1}\left(\frac{\sin(\omega T/2)}{(R + r_s)/R - \cos(\omega T/2)}\right) \qquad (3.1)$$

where ω is the angular velocity of rotation of the earth, T is the satellite orbital period, R is the earth radius and r_s is the satellite altitude. Figure 3.1 shows a plot of the scan angle, 2θ, needed for full coverage at the equator *versus* satellite altitude. The difficulty of achieving this type of coverage has prompted several investigators to propose four or more low earth orbiters to achieve the necessary coverage [2–3].

A question of interest in the statistical characterization of rainfall is the temporal resolution of the measurements over a particular area. Figure 3.2 shows the results of partial visits to an area 600 km × 600 km, located at the intermediate latitude of 15° N, by an orbiter in a 30° inclination [4] employing a satellite sensor

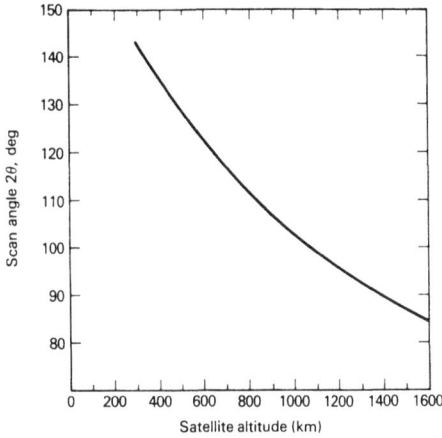

Figure 3.1 Total scan (2θ) needed for full coverage at the equator *versus* satellite altitude (from Okamoto *et al.*, [1]).

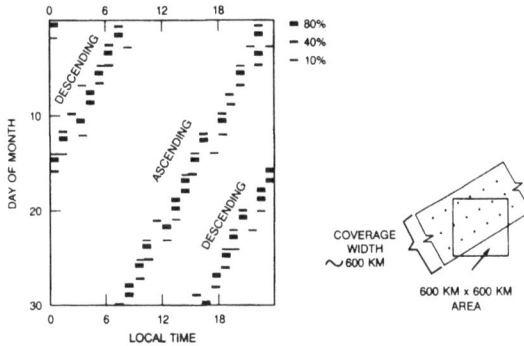

Figure 3.2 Visits to a 600 km × 600 km area at 15° N for an orbital inclination of 30° over a one-month period (from Simpson [4]).

with a swath of 600 km over an area. Shown in the figure are the partial visits to the area as functions of the day of the month and the local time. Although the number of partial passes over the area per month is about 80, this is equivalent to only about 30 full visits (that is, the fractional areal coverage per pass is on average ⅜ of the 600 km square). The results also indicate that over the course of a month the passes are fairly evenly distributed over the 24-hour period. This uniform sampling is useful in detecting diurnal patterns of rainfall on seasonal or annual time scales. Although polar orbiters with near sunsynchronous sampling lack this feature, it has been noted that a combination of a polar and an equitorial orbiter would improve the sampling accuracy substantially [4]. Perhaps the largest effect of the orbit on the radar design is the altitude variations of the spacecraft. If these variations are unknown, then the timing requirements and the PRF selection must reflect this uncertainty by including the appropriate margins in the interpulse period. The problem is more severe when transmission and reception are *staggered* and a common antenna is used than in a *burst-mode operation* or when separate transmitting and receiving antennas are used. For large uncertainties in the altitude, it may be necessary to adjust the PRF and timing based on the reception of the surface return spike at nadir—that is, to use the radar as a crude altimeter.

Typically, the choice of frequency is dictated by a combination of the hydrometeors of interest (cloud, light or heavy rain, snow) and a compromise among the desired resolution, swath, spacecraft altitude, and antenna size. Cloud sensing or simple rain detection capabilities favor the use of higher frequencies (near 35 GHz) and in the atmospheric window between about 75 GHz to 100 GHz. Making measurements over the full rain column (especially at the higher rain rates) requires a decrease in the frequency. Lowering the frequency presents certain problems, however, in that the ratio of the surface clutter to the rain signal increases (see Section 2.2), the sensitivity to light rain rates decreases (see Section 3.2), and larger antennas are needed to achieve the desired resolution. We note, however, that the degraded sensitivity is offset somewhat by the lower noise figures and more efficient transmitters that are available at the lower frequencies. Upgrading the radar from a single to a dual wavelength system has several advantages. The high frequency is best employed in detecting light rain rates and clouds, while the lower frequency (because of the smaller attenuation) is better able to sense moderate and heavy rains near the surface. The presence of returns from two wavelengths also permits the application of a number of estimation techniques that may lead to more accurate rain rate estimation. The dual-wavelength radar also enhances the capability for distinguishing the phase states of the hydrometeors (i.e., separating regions of rain from hail, the melting layer, and snow).

3.2 FREQUENCY SELECTION

Because the selection of frequency is discussed in other sections of this book (i.e., Sections 2.2, 3.5, and 5.3), we present a summary here of some of the more important considerations. Table 3.1 lists the frequency bands that have been allocated for space research and earth exploration satellites as set down by the International Telecommunications Union. Although petitions can be submitted for the use of any of these frequency allocations, the band from 35.5 to 35.6 GHz has been designated for spaceborne weather radars on a primary basis. To narrow the selection process, we note that for Rayleigh scattering and negligible attenua-

Table 3.1 Possible Frequency Bands for Spaceborne Weather Radar (ITU Radio Regulations)

International Frequency Band (GHz)	Primary Usage	Remarks
1.215– 1.300	Radiolocation[1]	1.28 GHz: SARs
1.525– 1.535	Space operation Earth Exploration S/C	(SEASAT-SAR, SIRs)
3.100– 3.300	Radiolocation[1]	
5.250– 5.350	Radiolocation[1]	5.3 GHz: SARs
8.025– 8.400	Fixed, Fixed S/C, Earth Exploration S/C	(ERS-1, SIR-C, D), Scatterometers
8.400– 8.500	Fixed, Mobile, Space Res.	
8.550– 8.650	Radiolocation[1]	9.6 GHz: SARs
9.500– 9.800	Radiolocation[1]	(SIR-D, X-SAR)
9.975–10.025	Radiolocation, Fixed[2]	
13.40 –14.00	Radiolocation, Space Res.[1]	13–15 GHz: Altimeters, Scatterometers
14.00 –14.25	Fixed S/C, Space Res.	
14.25 –14.30	Fixed S/C, Space Res.	
14.40 –14.47	Fixed, Space Res.	
14.50 –15.35	Fixed, Space Res.	
17.20 –17.30	Radiolocation, Earth Exploration S/C	
24.05 –24.15	Radiolocation, Earth Exploration S/C	
31.00 –31.30	Fixed, Space Res.	
31.80 –32.00	Radionavigation, Space Res.	
32.00 –32.30	Inter-S/C, Space Res.	
34.20 –35.20	Radiolocation, Space Res.	
35.50 –35.60	Spaceborne radar, Meteorological aids, Radiolocation	
65.00 –66.00	Earth Exploration S/C, Space Res.	
78.00 –79.00	Spaceborne radar, Radiolocation	

S/C: Satellite, Res.: Research

[1] Earth exploration satellite and space research services on a secondary basis.
[2] Allocated to the meteorological-satellite services on a secondary basis for use by weather radars.

tion, the frequency dependence of the backscattered power from hydrometeors, P_r, is contained in the term $f^{-2}G_0^2\eta$, where η is the radar reflectivity, f is the frequency, and G_0 is the antenna gain (see Section 2.1). To express this somewhat differently, note that G_0 is proportional to the physical area of the antenna, A_p, multiplied by the square of the frequency, and that for Rayleigh scattering the radar reflectivity is proportional to $|K|^2 f^4$. The frequency dependence of P_r, therefore, is represented by the quantity $A_p |K|^2 f^4$. Assuming for the moment that $|K|^2$ is independent of frequency, if the electrical size of the antenna, $A_p f^2$ is fixed, P_r increases as the square of the frequency. If A_p itself is constant, then P_r increases as the fourth power of the frequency. For example, if the physical size of the antenna is fixed and the frequency is raised from 3 GHz to 10 GHz, P_r increases by about 21 dB.

Despite the increase in sensitivity with frequency that the radar equation predicts, this must be balanced against the degradation in the performance of the radar transmitter and receiver. Figure 3.3 shows the current technology performance of high-power RF sources as of 1984 [5]. Figure 3.4 shows the frequency dependence of noise figures for *High Electron Mobility Transistor* (HEMT) amplifiers, which are considered to be the top-level performance figures as of 1987 [6]. As these data show, the noise figure of the receiver will increase with frequency, while the available power (and efficiency) will decrease. As a general guideline, we can say that for Rayleigh scattering, some sensitivity can be gained by using higher frequencies, although the degree of improvement will depend upon the system design and characteristics of the spacecraft hardware (see Section 3.5).

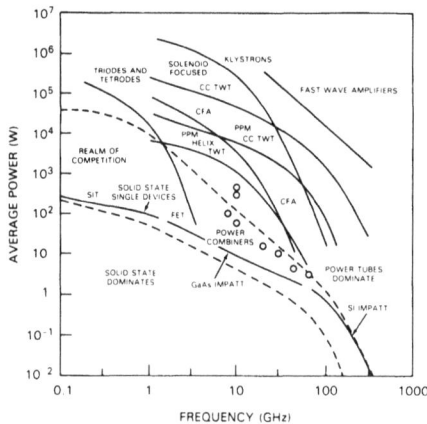

Figure 3.3 State of technology of RF sources as of 1984 (from Spielman [5]).

Above 10 GHz, $|K|^2$ generally decreases with increasing frequency (see Section 4.5); moreover, the backscattering from the larger particles no longer increases by the fourth power of frequency. Even more significant are the frequency effects of attenuation which, for large rain rates and penetration depths, can offset any advantage in sensitivity that the higher frequencies may afford. Figure 3.5 shows S/N ratios at the surface as a function of rain rate for a simple storm model

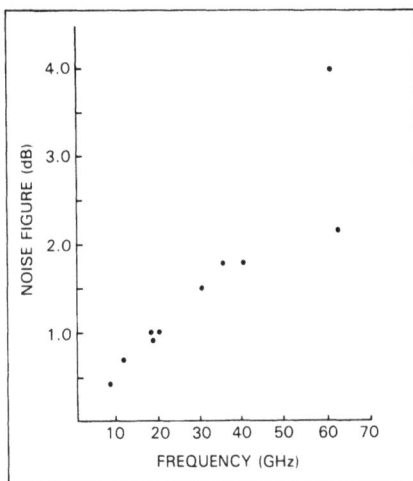

Figure 3.4 Noise figures for the HEMT amplifiers *versus* frequency (after Bierman [6]).

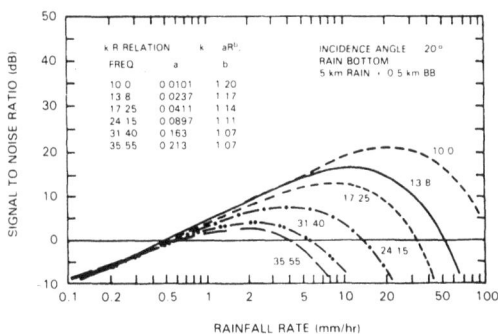

Figure 3.5 Rain signal to noise ratios (dB) *versus* rain rate for 6 radar frequencies at an incidence angle of 20°.

consisting of a 0.5 km melting layer, and a 5 km rain layer. For these results, the transmitted power has been adjusted so that the S/N ratios are approximately equal at the light rain rates. As the curves indicate, for frequencies above about 17 GHz and for penetration depths of 5 km or more, the dynamic range is severely limited by rain attenuation. Conversely, there are several drawbacks to the choice of a low-frequency radar. Apart from the loss in sensitivity at the light rain rates, attaining a resolution of 3 to 5 km for frequencies below about 5 GHz is impractical and costly. If a lower frequency is employed with poor resolution, the errors caused by surface clutter (see Section 2.2) and by reflectivity gradients within the beam will degrade the quality of the measurement. Thus, for rain sensing using a single frequency radar, the frequency band from about 10 GHz to 17 GHz offers a compromise between attaining adequate resolution and avoiding severe rain attenuation. In cases where the resolution is important or the size limitations on the antenna are severe, a higher frequency such as 35 GHz can be used for rain detection and estimation of precipitation parameters near the storm top. For a cloud radar, choices of 35 GHz or 94 GHz (or both) have been proposed [7–9]. Recent dual-frequency radar designs have studied the following combinations: (10 GHz and 24 GHz) [10–12]; (14 GHz and 35 GHz) [14–16]; (14 GHz and 24 GHz) [17–18].

3.3 SCANNING MODES

One of the most challenging aspects of spaceborne weather radar design is the need to sample a large volume of rain with sufficient accuracy and resolution in a time limited by the passage of the orbiter over the storm system. For the fan-beam meteorological radar, Katzenstein and Sullivan [19] suggested that the precipitation can be separated from the surface based on differences in the doppler spectrum. Such doppler discrimination, however, appears to be quite difficult unless the along-track beamwidth is very narrow. Moreover, for a radar operating at frequencies above 10 GHz and off-nadir incidence angles beyond about 30°, the rain signal should dominate the surface return, even at fairly light rain rates, so that a doppler mode for this purpose may be superfluous. In an extensive survey of weather radar designs, Skolnik [20] also considered the feasibility of the fan-beam approach. The geometry is illustrated in Figure 3.6.

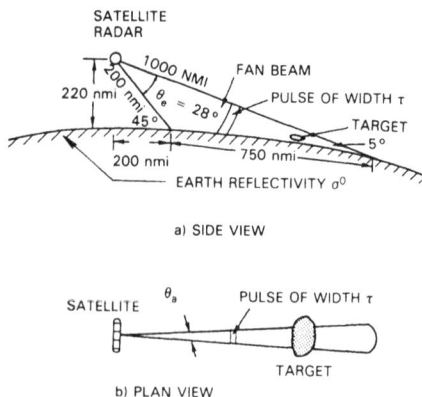

Figure 3.6 Fan-beam spaceborne geometry (from Skolnik [20]).

Some of the difficulties of the broadbeam approach include masking of the rain return by the surface clutter at near-nadir incidence, low S/N ratios resulting from the large cross-track beamwidth, and partial beamfilling of the resolution cell. One of the greatest drawbacks is poor vertical resolution. Consequently, large and unknown reflectivity gradients are probable. The presence of rain, ice, and partially melted hydrometeors within the pulse volume, moreover, will increase the difficulty for quantitative estimates of the precipitation properties. Despite these problems, some rain detection capability is possible. Skolnik [20] considered four radar designs with frequencies of 10 GHz, 16 GHz, 35 GHz and 94 GHz, each with azimuth and elevation beamwidths of 0.2° × 28°, off-nadir incidence angles ranging from 45° to 73°, a pulsewidth of 40 μs (6 km range resolution), a PRF of 80 Hz and average transmitter powers ranging from 500 W at 10 GHz (system noise temperature, T_{sys}, of 1000 K) to 50 W at 94 GHz (T_{sys} = 5000 K). For each of the designs, rain rates as low as 4 mm/h are detectable with S/N and signal-to-clutter (S/C) margins in excess of 10 dB. The effects of partial beamfilling and attenuation, however, do not seem to have been considered.

For most purposes, the fan beam is not particularly well suited to precipitation measurements. Nevertheless, any spaceborne sensor with rain detection capabilities should be quite useful. An analysis of the data from the *Shuttle Imaging Radar*, SIR-C, [21] should help determine the value of an auxiliary precipitation mode for the fan-beam antenna. The use of multiple beams, each with a separate transmitter and receiver, has been considered by Skolnik [20], Eckerman [22], and Okamoto [23]. In the study by Okamoto, the multiple beams are scanned through a portion of the swath (about 20 beamwidths) to provide contiguous coverage. The beam placement and the scan geometry are shown in Figure 3.7. The most common strategy by far is a real-aperture pencil beam employed in either a conical or cross-track scan. Illustrations of the cross-track and conical scan geometries are shown in Figures 3.8 and 3.9, respectively. To clarify some of the relationships between resolution, swathwidth and the number of independent samples, N, consider a cross-track scan between $\pm\theta_m$ at a fixed PRF, f_{PRF}. For continuous coverage, the available time per scan, T, is equal to the ratio of the along-track FOV to the spacecraft velocity, v_s. Letting H be the spacecraft altitude and θ_a, θ_c, the along-track and cross-track beamwidths, respectively, then

$$T = H\theta_a/v_s \tag{3.2}$$

For a uniform scan and a resolution cell defined by $\theta_a \times \theta_c$, the dwell time at each resolution cell, t, is approximately

$$t = H\theta_a\theta_c/2\theta_m v_s \tag{3.3}$$

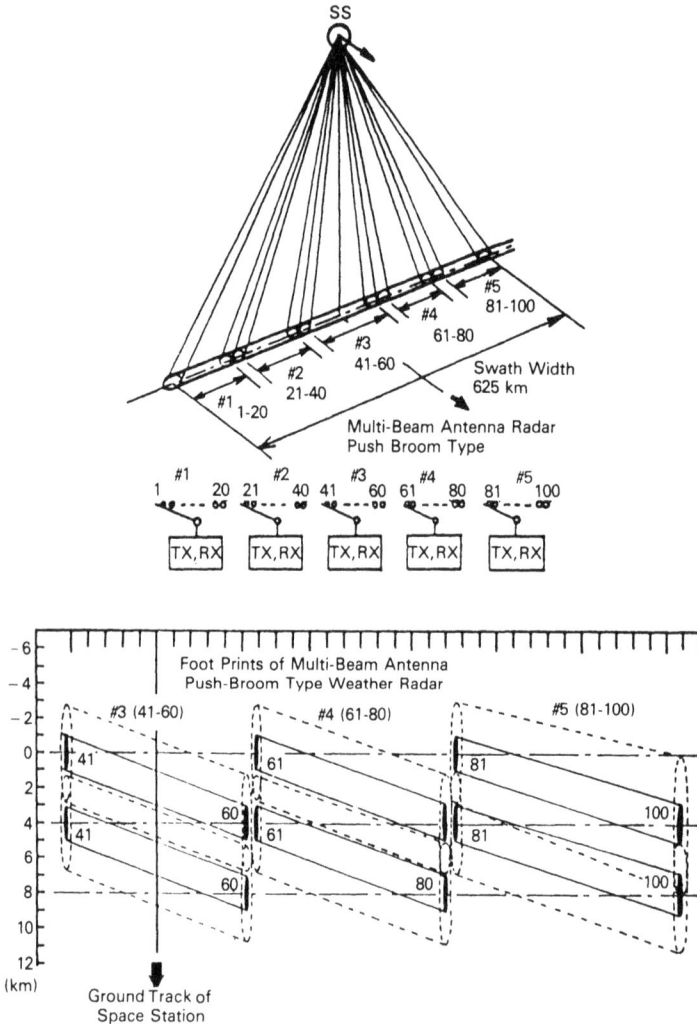

Figure 3.7 Illustration of a "push-broom" multibeam radar with subswath scanning (from Okamoto [23]).

We note that uniform sampling in angle corresponds to nonuniform sampling in space; i.e., the resolution cell increases because of the lengthening of the slant range. For a mechanically scanned antenna, the dimensions of the pulse volume in the range, along-track and "cross-track" directions* are approximately $[c\tau/2, (H - z)\theta_a \sec \theta$, and $(H - z)\theta_c \sec \theta]$ where τ is the pulse duration, θ is the scan

---

* In this context, the cross-track direction is given by $r \times v$, where $\hat{r}$ and $\hat{v}$ are unit vectors in the range and along track directions.

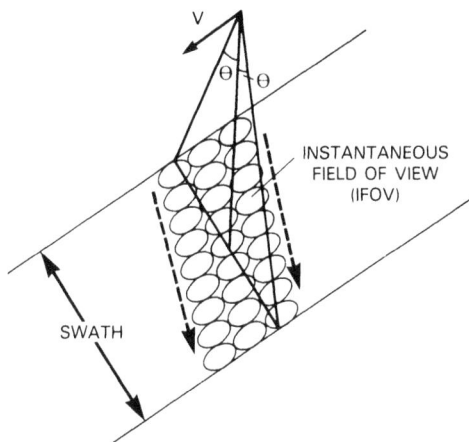

Figure 3.8 Geometry of a cross-track scan.

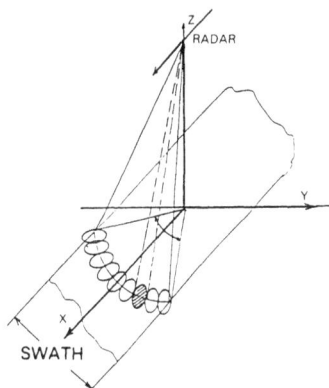

Figure 3.9 Geometry of a conical scan (Courtesy of R.K. Moore).

angle and z is the height of the pulse volume above the surface. For an electronically scanned antenna, because of the reduction in the projected antenna area for off-nadir angles, the dimensions of the pulse volume are approximately $[c\tau/2, (H - z)\theta_a \sec \theta$, and $(H - z)\theta_c \sec^2 \theta]$.

The number of samples per resolution cell, N, is given by the product of t and f_{PRF}. Approximating the antenna gain, G_0, by $\pi^2/\theta_a\theta_c$ yields

$$N = \pi^2 H f_{PRF}/2G_0 v_s \theta_m \tag{3.4}$$

For a fixed altitude and swathwidth $2H \tan \theta_m$ (flat earth approximation), N is inversely proportional to the product of the antenna gain and maximum scan angle, and directly proportional to the product of the altitude and the PRF. However, because the maximum PRF is a function of a number of parameters including θ_m, H, and the maximum storm height, H_s (see Section 3.4), the situation is somewhat more complicated. If we assume that the PRF is constant along the scan, then in order to prevent overlaps between successive returns, the PRF must satisfy the inequality:

$$f_{PRF} \leq 0.5c/[H(\sec \theta_m - 1) + H_s] \tag{3.5}$$

By substituting (3.5) into (3.4), the number of samples per resolution cell for a cross-track scan with a constant PRF satisfies the inequality:

$$N \leq \pi^2 c/[4\theta_m G_0 v_s(\sec \theta_m + H_s/H - 1)] \tag{3.6}$$

Curves of N $versus$ θ_m are shown in Figure 3.10 for $H_s = 20$ km, $H = 300$ km (solid lines) and $H = 1000$ km (dashed lines) for five values of the beamwidth θ_B, where $G_0 = (\pi/\theta_B)^2$.

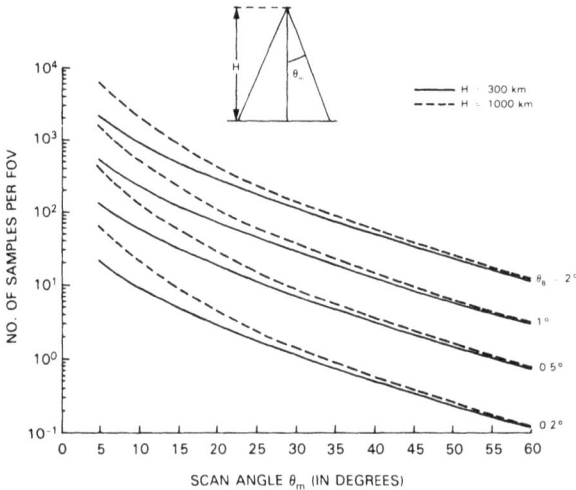

Figure 3.10 Number of samples per FOV for a cross-track scan and fixed PRF *versus* scan angle for satellite altitudes of 300 km and 1000 km.

If conical scan is used with $\theta = \theta_0$, the upper bound on the PRF is

$$f_{PRF} \leq c \cos \theta_0/2H_s \tag{3.7}$$

In this case, N satisfies the inequality:

$$N \leq \pi H c/(4H_s v_s G_0 \cos \theta_0 \sin \theta_0) \tag{3.8}$$

Equation (3.8) is obtained by recognizing that the antenna must complete a 360° scan in the time the spacecraft takes to move a distance $\theta_a H \sec^2 \theta_0$, and that the fraction of this time spent at each FOV is approximately $2\pi H \tan \theta_0/(H\theta_c \sec \theta_0)$. Other definitions of continuous coverage will result in slightly different expressions for N. For example, if we define continuous coverage with respect to the projection of the beamwidth onto the horizontal plane, then a complete scan must be made within a time $\theta_a H$, and the resulting N will be smaller than (3.8) by a factor of $\cos^2 \theta_0$. Further refinements in the definition can be made by including the effects of the earth's curvature, and by using the notion of effective horizontal resolution (see Section 2.1).

One of the advantages of the cross-track scan is that as the beam sweeps through nadir, the optimum vertical resolution is achieved. Moreover, at angles close to nadir, the fraction of the rain column over which significant surface clutter occurs is small. Some care is needed, however, to control the antenna sidelobe levels, especially in the range from about 5° to 20°. Moreover, in the case of pulse compression, the surface return from the range sidelobes is most severe at near-vertical incidence because of the large values of the backscattering cross section. Another disadvantage of the cross-track scan is that optimizing N requires that the PRF be varied with incidence angle (see Section 3.4). For conical scanning, the PRF can be fixed. The conical geometry also yields a stereoscopic viewing of the rain, and thereby lends itself to methods for measuring two-dimensional wind fields (see Section 2.8) or the rain attenuation (see Section 5.5). To a great extent, the scanning mode depends on the type of antenna. A mechanically scanned antenna is most suitable for conical scanning, while an electronic scan is best adapted for a cross-track scan (see Section 3.5). One of the design difficulties in a uniform scanning mode is the change in the antenna pointing direction in the time between transmission and reception of the signal. This problem becomes more severe as the scan rate and resolution increase. For electronic scanning, either the antenna incidence angle can be fixed over the full time needed for transmission and reception of the pulses, or the antenna can be dithered between the directions of transmission and reception during the transition from one angle to the next. For the more sophisticated adaptive viewing methods that are discussed below, electronic scanning is essential.

Adaptive Scanning

Over tropical regions, rain has been estimated to occur, on average, about 4 percent of the time [24]. This implies that for a preprogrammed scan, about 96 percent of the transmitted power will be expended over rain-free regions. The idea of *adaptive scanning* is to use the available time and power for sensing the regions of precipitation. Ideally, this scan strategy offers the potential for increased coverage, greater accuracy (higher dwell time) and power savings [3]. We begin by assuming that the available time per scan line, T, is fixed. If only a fraction of the swath contains the desired target, or if, for the particular application, certain rain intensities are more important than others, then several options are available: sample the rainy region for a time fT, where f is the fraction of beams containing rain out of the total number of beam positions per scan; use the entire time T to obtain a greater number of samples over those beam positions which contain precipitation; or expand the swath and allot a maximum number of beam positions, K, per swath, where K is smaller than the number needed for full coverage over the swath. For the first option, the demands on the satellite power are reduced, while in the second and third options, the additional time is used either to increase the number of samples or to increase the coverage.

For the options just mentioned, we assume that the full scan period can be used. This is true, however, only if some other instrument first defines the regions of precipitation. In the study by Atlas [3], a radiometer was proposed as the "look-ahead" sensor. Subsequently, researchers noticed that the radar itself could be employed in a rapid scan mode not only to locate the regions of rain, but also to perform a crude classification of the intensity levels [10–11]. Subsequent scanning would be conditioned on the importance of the various rain rate categories and the extent of the raining areas. For the TRMM radar, several adaptive scanning techniques have been analyzed, among which is a more modest implementation, where an initial rapid scan over the scene is not needed. Instead, a decision would be made after the reception of the first few pulses as to whether hydrometeors are present within the FOV. If the threshold is exceeded, signaling the presence of rain, then the remaining allotment of samples for that viewing angle would be transmitted; if the return is below the threshold, further transmission would be suspended until initiation of sampling for the next resolution cell. Apart from a much simpler implementation, this mode would provide continuous surface backscattering information, as well as some power savings. On the other hand, this design cannot be used to increase the radar swath or increase the number of samples. Studies of this and other adaptive scanning techniques have been made from the standpoints of system complexity and power savings [25].

3.4 TIMING CONSIDERATIONS

For a monochromatic radar, range ambiguities occur when the echo powers from successive transmitted pulses overlap. Thus, one of the requirements in determining the timing and PRF, f_{PRF}, is that only a single backscattered waveform be present at the receiver at any given time. For the following discussion, we assume that f_{PRF} is constant and that a cross-track scan is used. The minimum and maximum off-nadir scan angles are denoted by θ_{min} and θ_{max}. The minimum time delay between pulse transmission and the onset of the backscattered return waveform can be written as:

$$T_{min} = (2/c)(H - \Delta H/2 - H_s) \sec \psi_1 \tag{3.9}$$

with

$$\psi_1 = \max[0, (\theta_{min} - \theta_B/2 - \theta_e/2)] \tag{3.10}$$

where c is the speed of light, H_s is the storm height, and θ_B is the beamwidth in the cross-track direction. The quantity $\pm\Delta H/2$ represents the altitude variations about the nominal spacecraft altitude H, while $\pm\theta_e/2$ represents pointing errors in the incidence angle [26].

Equation (3.9) represents the minimum time required for the transmitted pulse to travel from the radar to the storm top and return. If ψ_1 is large or H_s is small, the surface return along the antenna sidelobes may arrive prior to this component. In the following discussion, we assume that this contribution can be neglected; in the actual system design, however, the amplitude and time of arrival of this interference term should not be overlooked. To find the maximum time delay between pulse transmission and the trailing edge of the echo, we note that the MI return (see Section 2.3) effectively increases the extent of the scattering medium. In formulating the equations for maximum delay, we must distinguish between the portion of the MI that we wish to record and the remaining portion that can be considered as interference [26]. To treat the general case, let $\tilde{H}_M$ be the height of the MI return over which the radar return is nonnegligible, with $\tilde{T}_{max}$ the corresponding time delay (i.e., the time delay between the pulse transmission and the trailing edge of the MI echo), and H_M the height of the MI over which the signal is to be recorded with T_{max} the associated time delay. Note that $H_M \leq \tilde{H}_M$. For a delta function transmitted pulse, T_{max}, and $\tilde{T}_{max}$ can be expressed as:

$$T_{max} = (2/c)[H + \Delta H/2 + H_M] \sec \psi_2 \tag{3.11}$$

$$\tilde{T}_{max} = (2/c)[H + \Delta H/2 + \tilde{H}_M] \sec \psi_2 \tag{3.12}$$

with

$$\psi_2 = \theta_{max} + \theta_B/2 + \theta_e/2 \tag{3.13}$$

Equations (3.9) through (3.13) can be applied to a conical scan by setting $\theta_{min} = \theta_{max} = \theta_0$.

Additional constraints on the timing and PRF must be imposed if the transmitted and received waveforms are "staggered" (that is, a burst mode is not used) and if the same antenna is used both for transmission and reception. These conditions can be stated in the following form:

(1) The backscattered waveform should be centered in the time "window" between adjacent transmitted pulses. In particular, the return from the first pulse is to arrive at the center of the interval between the nth and $(n + 1)$th pulse transmissions (Figure 3.11).

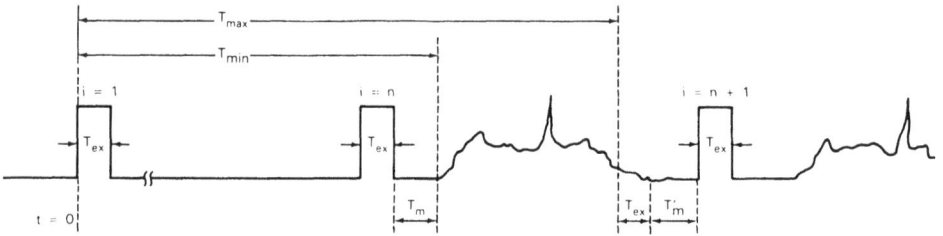

Figure 3.11 Timing requirements for the echo from the transmit pulse ($i = 1$) to arrive in the interval between the $i = n$ and $i = n + 1$ transmit pulses (after Awaka et al., [26]).

(2) The duration of the backscattered waveform, T_{ret}, must be less than the quantity $(T_p - \tau)$, where T_p is the interpulse period, $(1/f_{PRF})$, and τ is the duration of the transmitted pulse.

(3) Depending on the configuration of the circuitry, additional time margins may be needed for switching between transmission and reception, T_m, and from reception back to transmission, T'_m. The times required for changing the antenna pointing angle between directions of transmission and reception or for making measurements of noise power before the arrival of the backscattered power should be included in these margins.

(4) The MI return from the nth pulse must not overlap the return from the $(n + 1)$th pulse. This condition must hold in all cases (e.g., burst mode, separate receiving and transmitting antennas) unless frequency agility is used from pulse to pulse.

Conditions (1) through (4) can be represented by the equations:

$$f^{-1}_{\text{PRF}}(n - 0.5) = (T_{\max} + T_{\min} + T'_m - T_m)/2 \tag{3.14}$$

$$f^{-1}_{\text{PRF}} \geq (T_{\max} + T_{\min} + 2\tau + T'_m + T_m) \tag{3.15}$$

$$f_{\text{PRF}} \leq c \cos \psi_2/[2(\bar{H}_M + H_s)] \tag{3.16}$$

Assuming that $T_m = T'_m$, and substituting (3.9) and (3.11) into (3.14) and (3.15) gives

$$f_{\text{PRF}} = c(n - 0.5)/(H_{\max} \sec \psi_2 + H_{\min} \sec \psi_1) \tag{3.17}$$

$$f_{\text{PRF}} \leq 0.5c/[H_{\max} \sec \psi_2 - H_{\min} \sec \psi_1 + c(T_m + \tau)] \tag{3.18}$$

where

$$H_{\max} = H + \Delta H/2 + H_M \tag{3.19}$$

and

$$H_{\min} = H - \Delta H/2 - H_s \tag{3.20}$$

Because n must be a positive integer greater than 1, there exists a finite number of solutions of (3.16) through (3.18) for f_{PRF}. For the samples to be statistically independent (see Section 2.5), one other condition is imposed: the interpulse period must be greater than the time taken for the spacecraft to move approximately one-half the length of the along-track antenna dimension, L:

$$f_{\text{PRF}} < 2v_s/L \tag{3.21}$$

where v_s is the satellite speed. The largest value of PRF that satisfies (3.16) through (3.21) maximizes the number of independent samples per unit time.

Table 3.2 gives values of the PRF for the TRMM radar design [26]. The table is divided into three parts, corresponding to uncertainties in the satellite altitude of 5, 2.5, and 0 km. In the upper portion of the table, values of the PRF are shown for both *pulse compression* (PC) and *nonpulse compression* (NPC), under the condition that the PRF is constant throughout the full cross-track scan of $\pm 18.62°$. Because the duration of the pulse is typically much longer in the PC case, the PRF decreases. Nevertheless, the total number of samples at each resolution volume, N_d, can be made larger than the NPC case by summing the returns from consecutive groups of four range gates ($N_r - 4$) while retaining the same range resolution. For the NPC, comparisons are shown for cases where the PRF is fixed and where the PRF is allowed to vary (three values) as a function of incidence angle. By

Table 3.2 Number of Independent Samples for Several Values in the Uncertainty of Satellite Altitude [26]

(a) $\Delta H = 5$ km ($H = 320 \pm 2.5$ km)

	Scan Angle	PRF	N_r	N_d
Pulse-compression	$0° - 18.62°$	2560 Hz (Fixed PRF)	4	104
Non-pulse-compression	$0° - 18.62°$	3025 Hz (Fixed PRF)	1	31
	$0° - 9.31°$	4997 Hz	1	51
	$10.02° - 14.32°$	4447 Hz (Variable PRF)	1	45
	$15.04° - 18.62°$	4356 Hz	1	44

(b) $\Delta H = 2.5$ km ($H = 320 \pm 1.25$ km)

	Scan Angle	PRF	N_r	N_d
Non-pulse-compression	$0° - 18.62°$	3491 Hz (Fixed PRF)	1	35
	$0° - 9.31°$	5473 Hz	1	56
	$10.02° - 14.32°$	4916 Hz (Variable PRF)	1	50
	$15.04° - 18.62°$	4814 Hz	1	49

(c) $\Delta H = 0$ km ($H = 320 \pm 0.0$ km)

	Scan Angle	PRF	N_r	N_d
Non-pulse-compression	$0° - 18.62°$	3491 Hz (Fixed PRF)	1	35
	$0° - 9.31°$	5948 Hz	1	61
	$10.02° - 14.32°$	5384 Hz (Variable PRF)	1	55
	$15.04° - 18.62°$	5273 Hz	1	54

using a variable PRF, the number of independent samples can be increased by at least 50 percent on average over the number attainable with a fixed PRF.

3.5 SPACECRAFT HARDWARE CONSIDERATIONS

Performance demands on the radar system are created by the need to obtain accurate reflectivity measurements with good dynamic range in an environment with severe constraints on electric power, mass and size. In the following discussion, we focus on those aspects of the radar hardware that are relevant to spaceborne weather radar. Because of the importance of accurate rain reflectivity measurements, we also describe calibration techniques. A basic block diagram of a spaceborne weather radar is shown in Figure 3.12. The configuration of a spaceborne weather radar differs from that of ground-based radars in the presence of space-to-earth communication links for data transmission and for command or telemetry, and by the fact that the radar operation and data processing are closely related to the orbital altitude and attitude.

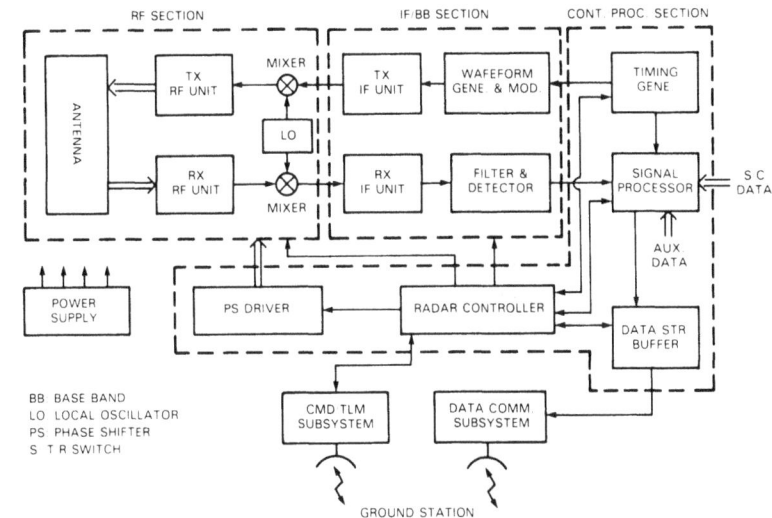

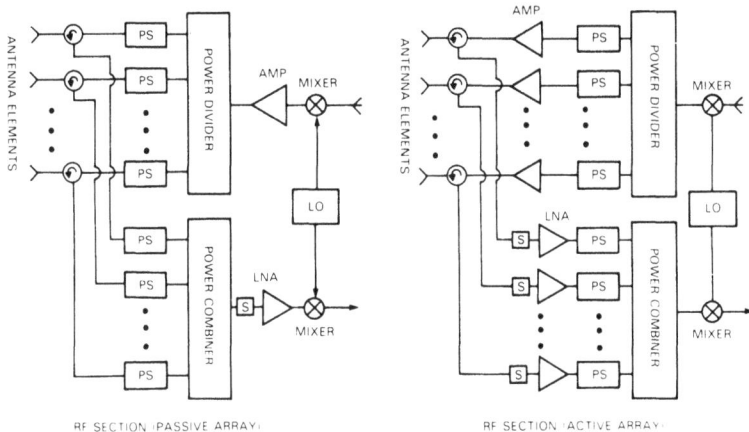

Figure 3.12 Top: basic block diagram of a spaceborne weather radar; Bottom: RF sections for passive and active array antennas.

Because the operation of spaceborne radars will be highly automated, computers will play a major role in the control and operation of the radar. Internal radar calibration circuitry is required to maintain and ensure accurate reflectivity measurements. Because the radar generates much greater heat than passive sensors and the space and environmental conditions are limited by the location of other instruments, thermal control is a significant concern.

Transmitter and Receiver

Choices for the transmitter include both vacuum tubes and solid state power amplifiers. Among the former are the *traveling wave tube* (TWT) and the klystron, both of which can satisfy the requirements of output power, efficiency, size, weight, and lifetime. The TWT, however, has been used for space applications far more often than klystrons, and a variety of space qualified TWTs are available.

Among the various performance specifications of the receiver, the noise figure is important in that an improvement in this quantity can lead to a substantial savings in the transmitter power while maintaining the same S/N. However, the extremely low noise temperatures attainable with devices such as the maser and the cooled parametric amplifier are not well suited for earth-viewing spaceborne radars, both because of the high background noise temperature and because the receiver tends to be massive with high power consumption. Semiconductor technologies are progressing toward devices with low noise temperatures while maintaining the desirable features of lightweight and small size [5–6]. One other important parameter of the receiver is its dynamic range. Due to the relatively small radar range variations over the target, the required dynamic range in rain is smaller than that required for typical ground-based radars. However, in order to measure the surface return over the ocean at nadir as well as the return from light rainfall requires a dynamic range on the order of 60 dB.

Antenna

There is no fundamental limitation on the choice of antenna type for a spaceborne weather radar as long as the antenna satisfies the requirements of gain, beamwidth, sidelobe level and scanning capabiiity. Among the spaceborne radars which have already been launched, nonscanning planar arrays have been used for most imaging radars and scatterometers, while nonscanning parabolic reflectors have been used for most radar altimeters [28].

1. Mechanical versus Electronical Scanning

The weather radar differs from other spaceborne radars in the requirement of wide coverage over a three-dimensional target. This has led most designs to employ some type of scanning. Mechanical scanning has been adopted in most spaceborne microwave radiometers because of the advantages of simplicity, low cost, a constant antenna pattern over the swath and a proven capability for accurate calibration. For a weather radar, however, the advantages of a cross-track scan (see Section 3.3) suggest the use of an electronic scan. This type of scan, moreover, affords greater flexibility in the manner and sequence in which the target is acquired; this is especially important in applications such as adaptive scanning. Conversely, the electronic scan generally implies greater cost and complexity, as well as difficulties in calibration due to changes in the beamwidth, and therefore in the radar "constant" with incidence angle.

2. Parabolic Reflector Antenna

The parabolic reflector antenna can be scanned either mechanically or electronically. In the latter mode, the electronically scanned feed system is usually combined with a fixed reflector. For a parabolic cylinder antenna, electronic scanning in the plane parallel to the cylindrical axis is possible by using a linear phased array as the primary radiator and by scanning its beam direction. For multifrequency or multibeam radar applications, an efficient use of space may require that several primary radiators use a single reflector. Because the primary sources cannot all be placed on the focal line, the shared use of a single reflector can be realized only at the expense of some degradation in antenna performance [29]. Figure 3.13 shows calculated antenna patterns of the offset parabolic cylinder antennas studied for the TRMM radar [27]. We can see that the displacement of

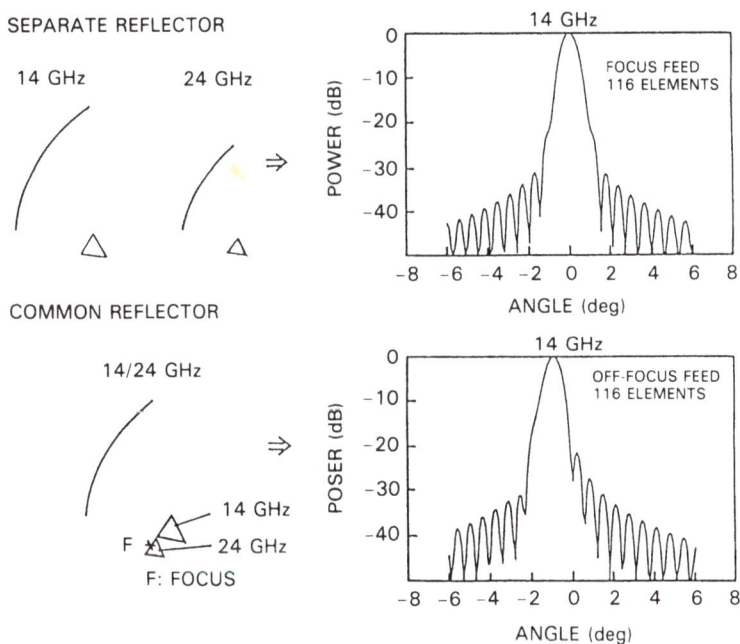

Figure 3.13 Degradation of antenna gain pattern for off-focus line feed (from Nakamura and Ihara [27]).

feeds from the focal line causes an asymmetry in the pattern and an increase in sidelobe levels.

3. Planar Array Antenna

Another type of antenna which is suitable for spaceborne weather radar is the planar array using either waveguides or microstrip circuitry. A versatile two-dimensional scan is possible if the phase of the RF signals provided to individual elements of the two-dimensional array are independently controlled. Although some ground-based tracking radars have taken advantage of this flexibility, the array requires high power and suffers from being heavy and costly. Moreover, for most applications, a one-dimensional electronic scan is sufficient.

One problem of phased array antennas is the existence of grating lobes. To avoid this problem, the interval between antenna elements, s, in plane of the scan should satisfy the condition:

$$s < \lambda/(1 + | \sin \theta_0|) \tag{3.22}$$

where λ is radar wavelength and θ_0 is the maximum scan angle. Because the actual antenna pattern is the product of the pattern of a single element and the array factor, the grating lobes appearing at large θ_0 may be reduced by modifying the former factor. Nevertheless, it is common to select an element interval which satisfies (3.22) to avoid the degradation of radiation properties. Phased array antennas can also generate multiple beams [31]. The active array, discussed below, is suitable for this application because the receiving antenna elements can be connected directly to the receiver front end, regardless of the number of beams. System loss, therefore, will not be degraded.

Passive and Active Array Systems

A phased array can be constructed in two ways. In the conventional *passive* array, the output signal from a single high-power transmitter is divided and sent to each antenna element. Conversely, the output signals from each antenna element are combined and sent to a single receiver front end. An alternative design is the distributed or *active* array, in which each antenna element has an associated transmitter and receiver front end [32]. In the active array radar, the RF signal from the driver is divided at the inputs of the final high-power amplifiers, while the received RF signals are combined after being amplified by the receiver front ends. Basic block diagrams of the RF sections of these systems are shown in the bottom portion of Figure 3.12.

The major features of the active and passive arrays are [27]:

(1) The passive array uses a single transmitter (final stage power amplifier) and receiver (front end), while the number of transmitters and receivers for the active array are equal to the number of antenna elements. Therefore, the passive array requires a high power transmitter, while the active array requires a large number of low power devices, which should be small and light so that the size and mass of the total system are kept within acceptable ranges.

(2) In the passive array system, phase shifters are inserted between the transmitter (or receiver) and the antenna elements. In the active array system, phase shifters can be inserted between the driver amplifier and the power amplifier (and between the front end and a frequency converter in the receiver). In the passive array system, therefore, the losses in the phase shifters are critical to the overall efficiency of the system. For active arrays, this is not the case.

(3) In the active array system, the length of feed lines connecting the transmitter (or receiver) to an antenna element can be quite short, thereby minimizing system loss. In the passive array system, however, this length tends to be long due to the distribution of the RF signal from the transmitter (and the combination of RF signals from the antenna).

(4) Because the RF signals are amplified independently, long-term phase and gain stabilities of each amplifier are critical to the performance and calibration of the active array system. Therefore, devices such as IMPATT amplifiers are not suitable for the active array radar due to their phase instabilities.

The selection of the transmitter and receiver is also closely linked to the choice of antenna. Because vacuum tubes such as the TWT and the klystron can provide sufficiently large peak powers, these devices are suitable for a passive array. On the other hand, solid-state devices are clearly more appropriate for the active array because of their compact size and the fact that integrated RF circuitry (including the transmitter, receiver and phase shifter) can be mass-produced. The type of phase shifter chosen also has a considerable effect on the system design [33]. Advantages of PIN diode phase shifters relative to ferrite phase shifters are: (1) light weight, small size, and compatibility with microwave integrated circuits; (2) short switching time (less than 1 μs); and (3) a power consumption that is independent of the phase switching rate (which provides the potential for complex scanning without an increase in power consumption). The ferrite phase shifter is generally heavy and large, with a switching speed that is low relative to PIN phase shifters. The device does, however, have the advantages of low insertion loss and the capability of handling high input powers. Therefore, ferrite phase shifters are suitable for passive arrays and PIN diode phase shifters are suitable for active arrays.

Pulse-Compression for Spaceborne Weather Radar

Pulse compression techniques eliminate the need for high peak powers and provide a greater number of independent samples by trading off resolution for independent samples. Despite these advantages, several problems exist, the most serious of which is the presence of range (time) sidelobes. Because the surface clutter (especially for near-nadir incidence) is usually much larger than the return from light rain, the range sidelobes can easily mask the rain echo. The reduction needed in the range sidelobe levels to minimize the interference from the surface return depends on the radar frequency, range resolution, and the rainfall rate (see Section 2.2). In general, the most stringent restrictions on the amplitudes of the range sidelobe levels are needed in the case of light rain rates at near-nadir incidence over an ocean background [25].

Weighting methods (both in the time and frequency domains) can be used to reduce the range sidelobes [34]. While it is theoretically possible to achieve very low range sidelobe levels, present performance is limited to sidelobe reductions of about 50 dB. Another possible problem is caused by signal decorrelation. This problem was first discussed in the use of pulse compression radar for ionospheric measurements [35], where Gray and Farley showed that the transmitted pulse length must be shorter than the decorrelation time of the scattered signal to avoid a degradation in range resolution and increases in the range sidelobe levels. A similar problem has been analyzed for the SAR azimuth compression [36].

Radar Calibration

System calibration is necessary to quantify radar reflectivity. Radar calibration can be classified into internal and external [37], where the former characterizes the system without including the antenna while the latter includes the properties of the antenna. *Internal calibration* can be further broken into two kinds [37]. In the first, each component is calibrated separately and the results are combined to obtain the total receiver gain. The alternative is the "ratio method," in which a known fraction of the transmitted signal is used as the reference signal to the receiver.

The meteorological radar equation relates the received power P (at the antenna output port) to the volume integral of the radar reflectivity weighted by the antenna gain, transmitted waveform, round-trip attenuation and range. However, the quantity that is measured is not P but the receiver output voltage, V. The transfer function, T, that converts V to P usually includes several nonlinear operations such as A/D conversion and logarithmic detection. Moreover, the receiver filter changes the transmitted waveform, and thus the range resolution. Because the final goal of radar calibration is to measure the scattering coefficient of targets as accurately as possible, we need not measure the system constant in the radar equation, C, and the transfer function, T, separately. The purpose of internal calibration is achieved if the product $C \cdot T$ is obtained. For this reason the ratio method, which determines the product directly, is preferable to calibrating each component. The basic block diagrams of internal calibration by the ratio method are given by Ulaby *et al.* [37].

From the perspective of total system calibration, *external calibration* is best because its objective is to relate V directly to the target scattering cross section by the use of a reference target. For point target calibration, there are several candidates, including corner reflectors and the Luneberg lens. As the typical horizontal resolution of spaceborne weather radars will be as large as several kilometers, a point target having a very large backscattering cross section must be used to obtain an adequate S/N. The active radar calibrator [38] is an attractive way to obtain a large cross section while maintaining a small target dimension. Extended reference targets are also free from S/N limitations. Because it is difficult to set up an extended artificial target over the radar scattering volume, natural targets such as a tropical rain forest have been used [39].

Because the radar equation varies depending on target type (point target, two-dimensional and three-dimensional extended targets), the three-dimensional extended target is obviously preferable for weather radar calibration. We can argue that the best target available for calibration of a weather radar is the rainfall itself. Typically the reference is obtained by converting the *rain rate*, as measured by surface rain gauges, to an effective radar reflectivity factor, Z_e, by means of a $Z_e - R$ relationship. This procedure, however, is subject to errors which arise from the widely different spatial and temporal measurement scales of the radar and rain gauges and from fluctuations in the drop size distribution. Selecting a uniform rain storm and estimating the $Z_e - R$ relationship from measurements of the drop size distribution diminishes these problems [40–42]. Another technique which uses the rain as reference target is based on adjusting the radar constant so that the path-integrated microwave attenuation derived from the reflectivity profile is equal to that measured either by a microwave radiometer or by the attenuation of a satellite beacon. This method has been tested successfully for ground-based radars in the estimation of path attenuation [41–43]. For downlooking observations from space, a similar approach could be taken by using the path attenuation as estimated from either a microwave radiometer or from the radar surface echo.

3.6 SYNTHETIC APERTURE RAIN RADAR

With the advent of the SIR-C and XSAR Shuttle Missions [21], interest in the use of SAR for precipitation sensing [44] is growing. Detection of rain using an airborne SAR was demonstrated over a decade ago [45]. To obtain an approximate expression for the S/N from a meteorological SAR, we start with the conventional result for the backscattered power from an extended target and replace the along-track beamwidth of the real aperture antenna, ϕ_0, with the synthesized beamwidth, ϕ_s. As the synthesized aperture is achieved by the coherent integration of n pulses, the return power, P_{sar}, is increased by the square of this factor. Assuming that the noise power is independent from pulse to pulse, the rms noise power, P_n, is equal to $nkT_{sys}B$ so that [44]

$$\frac{P_{sar}(r)}{P_n} = \frac{\lambda^2 G_0^2 n\phi_s\theta_0 c\tau\eta}{2^{10}\ln 2 r^2 (kT_{sys}B)} \exp(-0.2 \ln 10 \int_0^r k \, ds)$$

where θ_0 is the cross-track beamwidth, G_0 is the gain of the real aperture antenna, and η is the radar reflectivity.

Atlas and Moore [44] note that the quantity $n\phi_s$ is equal to the along-track beamwidth of the real aperture antenna, so the S/N for the SAR is equal to the ratio of a conventional radar having an along-track beamwidth equal to that of the real aperture SAR antenna. The use of SAR for meteorological targets has certain inherent limitations due to the motions of the hydrometeors. The mean radial velocity of the scatterers can result in a shift of the target outside the apparent resolution volume while doppler broadening caused by the motion of the hydrometeors ultimately limits the along-track resolution. Nevertheless, at least a three-fold improvement in resolution over the real aperture radar can be obtained from the SAR as long as [44,46]

$$\sigma_v < \lambda v_s/6L \tag{3.23}$$

where σ_v is the standard deviation of the doppler velocity spectrum due to turbulence, terminal velocity, shear, *et cetera*, but without the doppler beam-broadening term, and where λ is the wavelength, v_s is the spacecraft velocity, and L is the length of the along-track real aperture. To satisfy the Nyquist criterion f_{PRF} must exceed $2v_s/L$. If $v_s = 7$ km s^{-1}, then to satisfy (3.23) and the Nyquist criterion, L must satisfy the inequalities:

$$(1.17 \times 10^3\lambda/\sigma_v) > L > 14/f_{PRF} \tag{3.24}$$

where L is in meters, f_{PRF} in kHz, σ_v in m/s and λ in m. For example, if $f_{PRF} = 5$ kHz, $\lambda = 0.02$ m and $\sigma_v = 2$ m/s, then satisfying (3.24) requires that L be larger than 2.8 m and smaller than 11.7 m. If f_{PRF} can be doubled without range ambiguities, the lower limit on L can be reduced to 1.4 m.

Other difficulties for the meteorological SAR are coverage and sampling. Coverage can be attained with the usual large cross-track beamwidth, but at the expense of a degraded vertical resolution and an increase in interference from the surface. As the cross-track resolution is narrowed, scanning in the cross-track direction is required to broaden the coverage. This can be accomplished in a manner similar to the proposed EOS SCANSAR mode [47]. Independent samples can be obtained either by summing the returns from adjacent range gates using pulse compression or by spoiling the azimuthal resolution. Despite the problems with SAR for meteorological sensing, radars such as the SIR-C and XSAR may provide an important auxiliary precipitation mode. In addition, these radars should provide the first dual-polarization signatures of precipitation observed from space.

A closely related technique that should be mentioned is doppler beam sharpening, which has been considered as one of three modes of sensing for the proposed meteorological radar aboard the Space Shuttle [22,48–49]. System parameters for this radar are discussed in Section 3.7. For the beam-sharpening mode, the transmitting beam is very broad in azimuth (70°) and narrow in elevation. Although a receiving beam identical to the transmitting beam was originally considered, the large *time-bandwidth products* (where *time* refers to the interpulse period and *bandwidth* is associated with the doppler spread across the transmitted beam) imply that doppler processing would be highly ambiguous. To achieve time-bandwidth products of unity, 35 simultaneous receiving beams were proposed. Doppler processing on each of these 35 returns would be used to further sharpen the azimuthal resolution down to the scale of about 1 to 2 km.

3.7 SPECIFIC DESIGNS

As noted in Chapter 1, a number of spaceborne weather radar designs have been proposed since the 1960s. In this section, we concentrate on the more recent studies and proposals. We emphasize that many studies are concerned primarily with the science objectives, so the design issue is a secondary concern. Even in those programs where preliminary design studies have been completed, none is beyond a "Phase B" study. As such, the design parameters presented here are likely to undergo modifications; what should remain relatively fixed, however, are the basic design concepts as driven by the science requirements of the various missions.

Shuttle Meteorological Radar

Pioneering work in the design of a spaceborne weather radar was done by Eckerman [22], and Bucknam *et al.* [48–49] for a proposed Space Shuttle experiment. To attain broad coverage with adequate sampling, an *X-band* radar with multiple beams and receivers was adopted. A schematic of its three-dimensional storm mapping capabilities is shown in Figure 3.14. The radar was designed to accommodate three modes: real aperture, real aperture with pulsed doppler, and a azimuthal beam-sharpening mode. A block diagram of the radar is shown in Figure 3.15. The RF section can be used for all three modes. The objective of the *real aperture mode* is to obtain rain reflectivity measurements of high accuracy over a swath width of several hundred kilometers. The *real aperture-pulsed doppler mode* is used to provide an estimate of mean doppler shift, although the spread of the doppler spectrum places great demands on the processing. The major system parameters of the real aperture mode are listed in Table 3.3. To achieve a finer horizontal resolution, the *synthetic azimuth processing mode* is used. The radar waveform for this mode differs from the others and consists of repetitive bursts of pulses. After each burst is transmitted, the radar receives the return signals for all pulses within the burst. The burst duration is about five milliseconds, and the series of returns from each burst is doppler processed to yield an azimuth resolution of 1–2 km for most azimuth angles under moderate storm conditions.

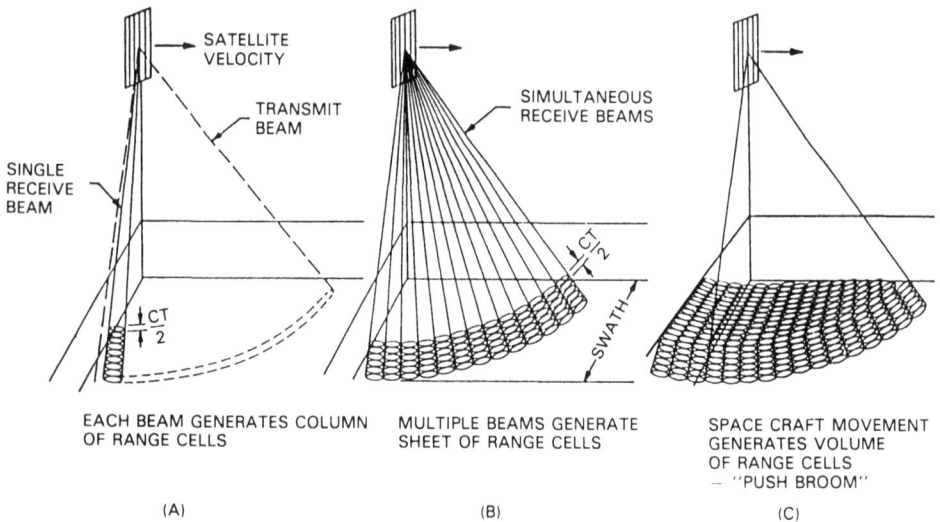

Figure 3.14 Concept of the 3-D rain mapping with a multibeam weather radar (from Eckerman [22]).

Millimeter-wave Radar

As noted in Section 3.2, an increase in the radar frequency yields the advantages of increasing the sensitivity for the detection of light rain and cloud, and higher angular resolution for a given antenna aperture size. Although the rain attenuation

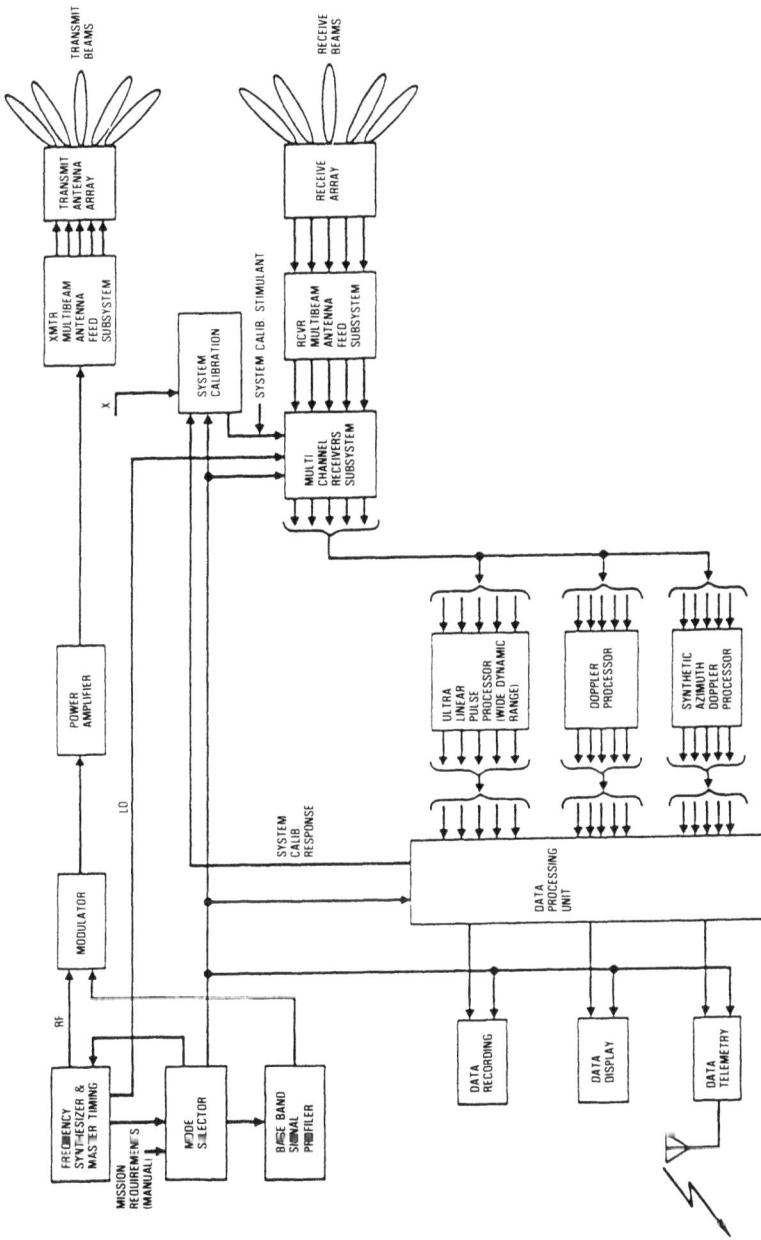

Figure 3.15 Block diagram of the proposed meteorological radar for the Space Shuttle (from Eckerman [22]).

Table 3.3 Design Parameters of the Proposed Shuttle Radar [22]

Shuttle altitude	200 km
Frequency	10 GHz
PRF	5.3 kHz
Peak transmit power	7.1 kW
Average transmitting power	250 W
Pulsewidth	6.7 μs
	(Range res. = 1 km)
Incidence angle	45°
Number of beams	73
Antenna dimension	5 m (H) × 4 m (W)
Swathwidth (each side track)	200 km
Horizontal resolution	3–4 km
	(Taylor weighting)

increases with frequency up to about 100 GHz, beyond about 30 GHz the increase is gradual. Consequently, the transmission "windows" between gaseous absorption lines (e.g., frequency ranges centered at 35 GHz, 94 GHz, and 140 GHz) can be used to detect rain or cloud [7–9, 50–51]. Lhermitte [7] proposed a spaceborne millimeter-wave radar operating at 35 and 94 GHz to detect cloud, light rain, and the bright band. Integrated atmospheric attenuation can also be estimated from sea surface echo levels. Assuming a satellite altitude of 800 km, an antenna diameter of 1.5 m, a pulse length of 2 μs, a receiver noise figure of 8 dB and a number of independent samples equal to 50, the 94-GHz radar can detect the tops of practically all clouds, including fair weather cumulus. However, the 35-GHz radar can detect only those clouds where the particle sizes extend up to about 100 μm [7].

Millimeter-wave Doppler Radar

Apart from the doppler nadir mode on the proposed TRAMAR, recent designs have focused almost exclusively on conventional pulsed radars. There are good reasons for this, as both large doppler shifts and spectral broadening introduced by the satellite motion place demands on antenna size, platform stability, and pulse repetition rate. By using a wavelength in the millimeter range, however, some of the difficulties may be less intractable than they would at first appear. In a recent study of spaceborne doppler weather radar, particular consideration has been given to a 94 GHz frequency [8]. This choice of frequency, along with a novel scanning strategy, is considered as a means by which two components of the wind field can be measured.

One of the most significant advantages of the 94 GHz radar is that an antenna dimension on the order of 2 m yields a beamwidth narrow enough that doppler beam broadening is reduced to an acceptable level. A second advantage arises from the fact that in the Rayleigh region (wavelength less than about 0.3 × particle diameter) the radar reflectivity increases as the fourth power of frequency. This implies that if the physical size of the antenna is fixed, the return power increases as the fourth power of frequency, while if the electrical size of the antenna is fixed, the return power increases as the square of the frequency. These relationships imply that a 94-GHz radar may have the capability of detecting a large fraction of the global cloud cover with a relatively modest transmitting power and antenna dimension.

Table 3.4 shows a comparison of three doppler radars at frequencies of 15, 35, and 94 GHz [8]. The radar parameters have been chosen such that all three designs provide approximately the same sensitivity to weak (Rayleigh) targets. A comparison between the 94 GHz and 15 GHz designs shows that the former design achieves the same sensitivity at the cloud top but with five-fold reductions in both the transmitted power and antenna diameter.

Table 3.4 Comparison of Design Parameters for Spaceborne Doppler Radar at Three Frequencies [8]

Frequency	94 GHz	35 GHz	15 GHz
Wavelength	3.2 mm	0.85 cm	2 cm
Peak power	1 kW	2 kW	5 kW
Average power (watts)	10	20	50
Pulsewidth	1 μs	1 μs	1 μs
PRF	10 kHz	10 kHz	10 kHz
Antenna diameter	1.8 m	5 m	10 m
Antenna beamwidth (degree)	0.1°	0.1°	0.12°
Footprint at 300 km (km)	0.5	0.5	0.6
Receiver Noise (dBm)	−104	−107	−109
Minimum detectable η (dB/cm^{-1})	−100	−114	−126
Minimum dBZ	−25	−22	−23
10 mm/hr^{-1} (two-way) rain att., dB/km^{-1}	9.7	4	1.2

Although the Rayleigh assumption is valid for cloud droplets at 94 GHz, departures from Rayleigh scattering even at light rain rates can be significant. Moreover, in clouds with a high liquid water content, the attenuation will be so severe that the penetration depth into the cloud or rain will be restricted. Despite these drawbacks, the high frequency approach may be the optimum solution for a spaceborne doppler weather radar. It would also serve as an excellent comple-ment to a lower frequency channel by constituting a radar with the capability of measuring a broad range of hydrometeors from clouds to heavy rains. To understand the scanning strategy that has been proposed for the doppler millimeter-wave radar, consider first a simple cross-track scan. For a fixed value of the cross-track angle, imagine the antenna nodding in the along-track direction first forward, then backward by an angle ψ, where ψ is on the order of 30°. With a 50 ms dwell time per FOV (which would provide 500 samples per FOV at a PRF of 10 kHz), a 10 km spacing between adjacent FOV in the cross-track direction, and a cross-track scan of ±30°, a sampling grid in the horizontal plane of about 10 km × 10 km can be achieved. This set of grid points within the storm would provide observations of the same rain or cloud volume from two aspect angles. If the antenna pointing angle is known and controlled with sufficient accuracy, then two components of the wind field can be found at those range gates where the beams intersect. The continuity equation could then be used to help infer the remaining component of the wind field [8].

Radar Sounder

The principal objective of the radar sounder is the measurement of cloud-base heights. Secondary goals of the mission are: estimation of cloud-top heights, cloud water content, the spatial extent and intensity of precipitation, and the location of the melting layer [9,51].

In the selection of the radar frequency, those lower than Ka-band can be ruled out because of poor cloud detectability; beyond about 140 GHz, atmospheric, cloud and rain attenuation is prohibitive. Taking into account the limitations of the amplifiers at the higher frequencies, transmitter efficiencies, and antenna rms deviation tolerances, the Ka-band channel (35 GHz) was found to have better sensitivity than 13 GHz, 94 GHz, or 140 GHz not only for stratus and cirrus clouds but for light rain rates as well. Focusing on the Ka-band channel, three types of design were considered. The first is a reconsideration of the Space Shuttle design [48–49] which proposed using transmitting beam broad in azimuth and narrow in elevation with multiple receiving beams. Despite advances in the technology, the antenna is considered too large and costly. A second option is SAR: for the standard fixed beam antenna, coverage is insufficient, but use of the SCANSAR concept to broaden the swath [47] leads to a complex processor with no compelling advantage over a scanning pencil beam. The third and preferred option is a conically scanned paraboloid using a conventional (unchirped) pulse waveform. Because a narrow beam is desirable, the angular rotation of the antenna must be high, introducing concerns of momentum compensation and mechanical failure. To reduce the angular velocity, a dual-beam feed (common reflector) is proposed. A further decrease in the angular velocity is gained by noting that a 2-D spectral analysis of the rain suggests that a 6 km sampling density can completely characterize the spatial distribution. Thus, for a 2 km horizontal resolution and a 6 km spacing between adjacent FOV, the dual-beam rotation rate can be lowered to 17 rpm. Preliminary design parameters for the radar sounder for the very high resolution option are given in Table 3.5 [9].

Table 3.5 Design Parameters of the Radar Sounder for the Very High Resolution Case [9]

Frequency (radar)	35.6 ± .30	GHz
Frequency (radiometer)	36.6–37.6	GHz
Altitude of Satellite	800	km polar
Antenna Diameter	4.1	m
Beamwidths (one-way)	.15–.16	degrees
Horizontal Resolution	2	km
First Sidelobes (35 dB design)	31	dB
Gain (G_T, G_R) (40° efficiency)	59.7	dB
Number of Transmitters and Receivers	2	
Transmit positions	2	per transmitter
Peak Power (per transmitter)	6.2	kW
Average Power (per transmitter)	150	W
Pulse Repetition Frequency (nominal)	7	kHz
Pulse Duration	3–4	μs
Vertical Resolution	<1	km
Transmitter Loss (to feeds)	2.50	dB
Pattern Losses	1.0	dB
L_R, Receiver Loss (included in T_S)		
T_S, System Noise Temperature (L_R = 2.0 dB)	500	K
Filter and Detector Losses	1.0	dB
Distribution or CFAR Loss	1.0	dB
Swath Diameter	400, 600	km
Nadir Angles	13.6, 20	degrees
Rotation Rate (approx.)	17	rev/min
Pointing Accuracy (≈1 km)	.11	degrees

Modified Radar Altimeter

The use of existing spaceborne radars such as SARs, scatterometers, and altimeters for rain observations from space would substantially increase weather monitoring capabilities. The altimeter is perhaps the best choice for testing the feasibility of rain measurements from space, because nadir viewing allows for good vertical resolution of the storm structure without interference from the surface clutter. Furthermore, the band in which most altimeters operate (13–14 GHz) is well suited for rain observations. Goldhirsh and Walsh [52] showed that a minor modification of a SEASAT type of radar altimeter could provide estimates of rain rate. Table 3.6 lists the major parameters of the SEASAT altimeter and the proposed modifications needed for measurements of rain. The SEASAT altimeter transmits a 3.2 μs chirped pulse at a PRF of 1000 Hz. The received pulse is compressed to 3.125 ns to obtain a range resolution of 0.5 meters. Using the same 3.2 μs pulse without chirp modulation provides a range resolution of 480 meters. In this mode, the rainfall just above the surface could be observed with S/N larger than unity for rain rates ranging from about 1 mm/h to 55 mm/h. (See Figure 5.1, Chapter 5.) By operating in the rain mode about 5 percent of the time, the rain profile can be deduced from an average of 45 samples.

SIR-C

Although SAR is not designed for rainfall observations, the radar might provide such information if the sensitivity and resolution are sufficiently high. The imaging radars previously launched have used L-band ($\approx$1.3 GHz). Consequently, their

Table 3.6 Parameters of SEASAT Altimeter and those of Rain Measurement Mode [52]

Satellite altitude	800 km
Frequency	13.5 GHz
PRF	1000 Hz
Peak transmitting power	2 kW
Beam-limited FOV	22 km
Nominal (altimeter) mode	
Pulsewidth (before compression)	3.2 μs
Pulsewidth (after compression)	3.125 ns
Frequency deviation (bandwidth)	320 MHz
Pulse-limited FOV	1.7 km
Interval of range bin	0.5 m
Number of range bin	60
Rain measurement mode	
Pulsewidth	3.2 μs
Bandwidth	0.3 MHz
Interval of range bin	500 m
Number of range bin	60
Number of independent samples	45
Operation sequence	
Nominal mode	The first 19/20 of each second (950 pulse hits)
Rain measurement mode	The last 1/20 of each second (50 pulse hits)

sensitivity to rain has been poor. The *Shuttle Imaging Radar C* (SIR-C) experiment is scheduled to be launched in 1991. The sensors consist of a C-band (5.3 GHz) and L-band SAR. In addition, the Federal Republic of Germany will provide an X-band SAR. Although the broad cross-track footprint at C-band of about 17 km poses difficulties for any quantitative measurement of rain, the SIR-C will have electronic beam-steering capabilities. Thus, by employing low elevation angles for rain sensing, the degradation of the measurement due to poor vertical resolution can be offset to some extent. Table 3.7 lists the major performance parameters of the SIR-C [21]. According to Atlas and Moore [44], the rainfall rate at which the S/N is unity is about 1 mm/h, provided that the antenna beam is filled with rain.

Conical Scanning Radar

As described in Section 3.3, the conical scan differs in several respects from the cross-track scan. The constant incidence angle of the conical scan allows for

Table 3.7 Parameters for the SIR-C Shuttle Radar [21]

Parameters	SIR-C
Orbital altitude	225 km[a]
Orbital inclination	$\geq 57°$
Frequency	1.25 GHz[c]
	5.3 GHz[c]
	9.6 GHz[b]
Polarization	*HH* (L. C)
	VV (L. C, X)[b]
	VH (L, C)
	HV (L, C)
Incidence angle	15°–55°
Swathwidth	15–90 km[c]
Azimuth resolution	30 m (4 look)[c]
Range resolution	60–10 m[c]
Peak power	3.8 kW (L)
	2.1 kW (C)
	3.3 kW (X)[b]
Bandwidth	10 or 20 MHz
Optical data collection	—
Digital data collection	50 h/channel.
	5 channels

[a] Approximate
[b] *X*-band *VV*-polarized SAR to be added by DFVLR
[c] Based on preliminary design parameters

higher PRFs, and may simplify the data interpretation. If a narrow beamwidth can be attained, then the large off-nadir incidence angle tends to reduce the magnitude of the surface clutter. Moreover, in the case of pulse compression, the range sidelobe problem is much less severe than in a cross-track scan. Achieving adequate vertical resolution, however, requires a large aperture antenna. Parameters of a conical scanning radar for a 500 km spacecraft altitude are given in Table 3.8 [53].

Table 3.8 Parameters of a Conical-scan Rain Radar [53]

Satellite height (km)	500
Off-vertical angle (deg)	32
Swath width (km)	358
Beam footprint dimensions (km)	4 × 4.7
Number of independent samples	170
Peak transmitted power (W)	150
Range resolution (m)	30
Pulse repetition frequency (Hz)	7000
Receiver bandwidth (MHz)	5
Expanded pulse length (μs)	50
Compressed pulse length (μs)	0.20
Frequency (GHz)	15
Wavelength (cm)	2.0
Sensitivity (mm/hr) (for $S/N = 0$ dB)	
Rain top	0.35–100
Rain bottom	0.35–60
Antenna dimensions (m)	3 × 3
Beamwidth (°)	0.38
Antenna gain (dB)	54.5

TRMM Rain Radar

The major scientific goal of the *Tropical Rain Measuring Mission* (TRMM) is the measurement of the distribution of rainfall in the tropics (35° s to 35° N) for climatological studies and model verifications [4,23,54]. The major measurement objective is to obtain a three-year data set of monthly rainfall amounts over a grid 500 km on a side with a sampling error of less than 10 percent. A schematic of the TRMM concept is shown in Figure 3.16. A summary of the TRMM sensors is shown in Table 3.9 [4]. Although a dual-frequency radar was originally proposed, because of power (and budgetary) constraints the radar is now limited to a single frequency near 14 GHz. A higher frequency (24 GHz) channel will be added if resources permit. From an altitude of 350 km, the antenna, which will provide a

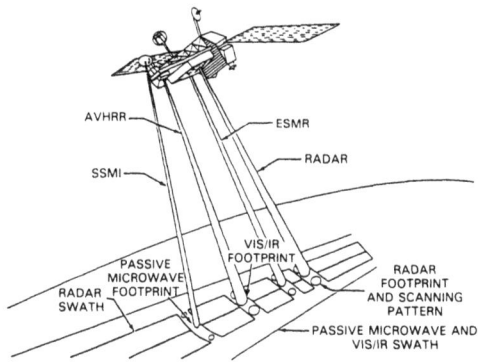

Figure 3.16 Illustration of rain observations from the TRMM satellite (from Simpson [4]).

Table 3.9 Summary of the Proposed TRMM Sensors [4]

Microwave Radiometers	Radar	Visible-Infrared Radiometer	Orbit
19, 37, 90 GHz (dual polarized)	14 GHz	VIS and 10 μm IR	35° inclination
10 km resolution	4 km footprint	1 km resolution	350 km altitude (high resolution)
600 km swath	250 m range resolution 220 km swath	600 km swath	Rapid precession
10 GHz at 20 km resolution	*24 GHz*	*Moonlight visible*	
5 GHz at 40 km resolution	*600 km swath*	*1.6 mm (phase of H$_2$O)* *6.7 mm, split window*	

Note: Items in italics desired if resources permit but are not necessary to achieve main TRMM objectives.

horizontal resolution at nadir of about 4 km, will be scanned electronically over a swathwidth of 220 km (Figure 3.16). A range resolution of 250 m is required at nadir to obtain vertical storm structure. For off-nadir angles, a degraded range resolution may be used to reduce the dwell time while maintaining an adequate number of independent samples. It was concluded that at least 64 independent samples would be required to achieve an adequate measurement precision.

Conceptual designs of the TRMM rain radar have been made in Japan [17–18,25–27,55–56] and in the US [13–14]. Several features of the TRMM radar, such as the determination of the PRF and scanning strategy have been mentioned in previous sections. Table 3.10(a) summarizes the main features of the six design options that have been studied; the corresponding radar design parameters are shown in Table 3.10(b) for the fourth option [18]. As indicated in Table 3.10(a), trade-off studies have been made with respect to the antenna (parabolic cylinder *versus* planar array and active *versus* passive systems), the waveform design (conventional *versus* pulse compression), and the transmitter (TWT *versus* SSPA). Although several candidates still remain, the conventional pulsed radar has been given a higher priority due to the range sidelobe problem inherent in the pulse compression system [18]. A complex scanning strategy (variable PRF and nonlinear scan) has been proposed to increase the number of independent samples. More sophisticated adaptive scanning techniques for both the TRMM and TRAMAR designs have been considered for either increasing the sampling density or widening the swath [10–13,25]. Figure 3.17 shows the S/N for the rain and sea surface echoes calculated using the radar parameters in Table 3.10(b), and a storm model which consists of uniform rain of 5-km depth and a melting layer of 0.5-km thickness [18]. We note that the differences in performance among the various options are minor.

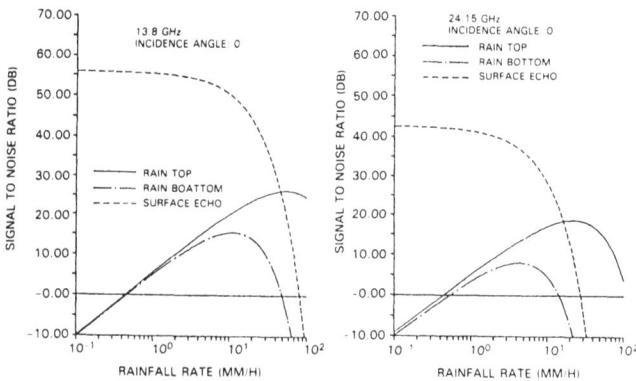

Figure 3.17 Rain S/N *versus* rain rate from the rain and sea surface for the proposed TRMM radar (from Okamoto *et al.*, [18]).

Space Station Rain Radar (TRAMAR)

As a follow-on to TRMM, the *Tropical Rain Mapping Radar* (TRAMAR) has been proposed for the *Earth Observing System* (EOS) on the Space Station [10–13]. Table 3.11 summarizes the major parameters of TRAMAR. The dual frequency radar (10 and 24 GHz) will be provided with a nadir-looking experimental doppler

Table 3.10(a) Summary of the Options under Study for TRMM Radar Design [18]

	Case 1	Case 2	Case 3	Case 4	Case 5	Case 6
1. Antenna	planar	planar	planar	planar	cylindrical	cylindrical
2. Transmitting Power Amp.	TWTA	TWTA	SSPA	SSPA	TWTA	TWTA
3. PC or NPC	PC	NPC	PC	NPC	PC	NPC
4. Power consumption	339 W	315 W (378 W[*1])	396 W	324 W (355 W[*1])	339 W	315 W (378 W[*1])
Margin	61 W	85 W (22 W[*1])	4 W	76 W (45 W[*1])	61 W	85 W (22 W[*1])
5. Weight	310 kg	310 kg	468 kg	560 kg	300 kg[*2]	300 kg[*2]
6. Size 13.8 GHz (mm)	2100×2100×460	2100×2100×460	2300×2300×650	2300×2300×650	3000×2520×1550[*2]	3000×2520×1550[*2]
24.15 GHz	1250×1250×300	1250×1250×380	1300×1300×650	1300×1300×650	3000×2520×1550[*2]	3000×2520×1550[*2]
7. ΔS/N 13.8 GHz	-1.0 dB	+0.6 dB	-1.5 dB	+2.1 dB	-1.0 dB	+0.6 dB
24.15 GHz	-0.9 dB	+0.8 dB	-1.4 dB	-0.7 dB	-0.9 dB	+0.8 dB

[*1]: When variable PRF and non-linear scan are used.
[*2]: Size of occupied volume by a cylindrical parabolic antenna system (reflector + feeder). This size assumes that only the feeder of 13.8 GHz is considered and increases if both 13.8 GHz and 24.15 GHz feeders are considered.

Table 3.10(b) TRMM Radar System Parameters for Option 4 [18]

	13.8 GHz	24.15 GHz
Antenna		
Type	Planar array	Planar array
Polarization	HH	HH
Gain	47.7 dB	47.7 dB
Beamwidth	0.716 x 0.716 deg.	0.716 x 0.716 deg.
Aperture	2.2 m x 2.2 m	1.2 m x 1.2 m
Sidelobe	less than −35 dB	less than −35 dB
Phase shifter (*1)	PIN-diode (5 bit)	PIN-diode (5 bit)
Number	131 × 2	131 × 2
Loss	−	−
Scan angle	±18.6 deg.	±18.6 deg.
Angle bin number	53	53
Scan period	0.545 s	0.545 s
Transmitter		
Type	SSPA (×131)	SSPA (×131)
Peak power	572.1 W	200.2 W
Efficiency (*2)	17.5 %	9.63 %
Pulse width	1.67μ s	1.67μ s
PRF (Fixed)	3025 Hz	3025 Hz
Duty (*3: Fixed PRF)	1.01 %	1.01 %
PRF (*4: Variable PRF)	4787 Hz	4787 Hz
Duty (*3, *4: Variable PRF)	1.60 %	1.60 %
Receiver		
Noise figure	2.5 dB	4.0 dB
IF frequency	160 MHz	160 MHz
Band width	0.78 MHz	0.78 MHz
Characteristics	Logarithmic	Logarithmic
Dynamic Range	More than 70 dB	More than 70 dB
Linearity	±0.5 dB	±0.5 dB
Total system loss	2.0 dB	2.5 dB
Number of independent samples (Fixed PRF)	31	31
Number of independent samples (Variable PRF + Non-linear Scan)	65 (PRF = 4997 Hz) 64 (PRF = 4447 Hz) 64 (PRF = 4356 Hz)	65 (PRF = 4997 Hz) 64 (PRF = 4447 Hz) 64 (PRF = 4356 Hz)
Data rate	170 Kbps	−

(*1): Separate phase shifters are used for transmitters and receivers.
(*2): Including power supply.
(*3): Including margin of 100 %.
(*4): Average value; PRF changes as scan angle changes.

Scan Angle	PRF
0 – 9.31 deg.	4997 Hz
10.02 – 14.32 deg.	4447 Hz
15.04 – 18.62 deg.	4356 Hz

mode that will be operated about one percent of the time. In the standard operating mode (nondoppler), an adaptive cross-track scan will provide a swathwidth of about 800 km, corresponding to a scan angle of about ±45° about nadir. An 11.5 m × 3.8 m flat, rectangular reflector antenna with a dual-frequency offset feed

Table 3.11 Science Requirements and System Parameters for TRAMAR [11]

(a) Science Requirements

A. Rain rate range:	from 0.5 to 100 mm/hr
B. Ground swath coverage:	800 km
C. Observable rain column height:	0 to 15 km above the surface
D. Instantaneous rain measurement qualities:	
Horizontal resolution:	4 km x 4 km
Vertical resolution:	250 m at nadir to 1.5 km at swath edges
Rain rate accuracy:	40% systematic; 50% random (0.5 – 2 mm/hr)
	20% systematic; 30% random (2 – 4 mm/hr)
	10% systematic; 20% random (4 – 100 mm/hr)
Rainfall velocity accuracy:	1 m/sec (for Doppler measurements only)
Location accuracy:	1 km
E. Monthly averaged rain measurement qualities:	
Horizontal resolution:	5° latitude by 5° longitude
Vertical resolution:	250 m at nadir to 1.5 km at swath edges
Rain rate accuracy:	15% (monthly average)
Location accuracy:	10% of the horizontal resolution cell size
F. Data timeliness(maximum time lag):	
Archived 3-D rainfall map:	1 month
3-D rainfall map for regional experiments:	5 hours

(b) System Parameters

	9.7	24.1
Frequency (GHz)	9.7	24.1
Polarization	VV	VV
Antenna		
Effective aperture (m)	11.6×3.8	11.6×3.8
Beamwidth (3-dB one way) (deg)	0.164×0.506	0.066×0.506
Peak gain at nadir (dB)	52.7	56.3
Peak side-lobe (dB)	-27	-27
Scan angle from nadir (deg)		
Normal Mode	±40 - ±50	±40 - ±50
Doppler Mode	0	-
Transmit peak power (kW)	2	1
PRF (KHz)		
Normal Mode	3.0 - 4.3	3.0 - 4.3
Doppler Mode	5.3 - 5.4	-
Pulse duration (μsec)		
Normal Mode	6.7	6.7
Doppler Mode	1.7	-
Frequency dither in Normal Mode	4	4
Bandwidth (MHz)	1.8	1.8
Transmit path loss (dB)	-1.5	-2.5
Receive path loss (dB)	-2.0	-3.0
Digitization noise (dB)	-0.5	-0.5
System noise temperature (deg)	758	1411
System dynamic range (dB)	68	60
Instantaneous data rate (kb/s)		
Normal Mode	80	80
Doppler Mode	9000	-
Averaged data rate (kb/s)	250	

will yield an FOV of about 4 km for both frequencies. The dipoles etched on the reflector surface, when illuminated by the feed, will be designed to simulate the fields on a cylindrical parabolic reflector. The range resolution of 250 m is the same as the TRMM radar. However, at off-nadir angles, a higher vertical resolution will be attained because of the narrow cross-track antenna beam. To achieve a greater number of independent samples, frequency agility will be employed. This is to be accomplished by dividing the total pulsewidth of 6.7 μs into four 1.7 μs segments, each with a frequency sufficiently separated from the others to obtain mutual statistical independence. The resulting number of independent samples per FOV is estimated to be about 100 [10–12].

BEST (Bilan Energetique du Systeme Tropical)

This instrument complement is designed for measuring the various components of the energy budget of the Tropics [15–16]. The primary objectives are similar to those of TRMM and TRAMAR, but in terms of the number of sensors, and in radar resolution BEST is more ambitious. The baseline set of instruments consists of a radar with either a single (14 GHz) or dual-frequency capability (14 GHz and 35 GHz), a conically scanned multichannel radiometer and a pulsed doppler CO_2 LIDAR at 9.1 nm for wind profiling. As in TRMM and TRAMAR, the proposed radar scan is cross-track, using a real aperture pencil beam. For adequate sampling of the rain, the goal is a horizontal $\times$ vertical resolution of 1.5 km $\times$ 0.25 km with at least 60 independent samples per FOV for a single frequency, and at least 140 for a dual frequency. The more stringent requirement for the dual-frequency radar arises from the greater sensitivity to fluctuations in dual-frequency methods, because the estimates are generally formed from a ratio of powers at the two frequencies. The upper and lower bounds on the swath are determined by setting the upper limit for a maximum off-nadir angle of 20° and a lower limit of 100 km. Technological limitations are now considered to be a peak transmitted power of 1 kW for a 1 μs pulse duration, and a maximum antenna size of 10 m on a side. Because of the narrow beamwidth and the desirability of a wide swath, an increase in the number of samples may be necessary. For this purpose the use of dual-mode frequency agility (13.75, 13.76 GHz) has been proposed as an option. Parameters of the nominal design are shown in Table 3.12 [16].

As noted in Section 3.3, one of the design issues for spaceborne weather radar is in satisfying the competing demands of wide coverage, high resolution, and adequate sampling. For BEST, a greater emphasis is placed upon high resolution measurements near nadir. By choosing a spatial sampling interval of 3 km (with an IFOV of 1.5 km), the design affords a relatively high sampling density near nadir and a reduction in reflectivity gradients across the beam. This high resolution and sampling density comes at some sacrifice in coverage and a signifi-

Table 3.12 System Parameters and Science Requirements for BEST [16]

System Parameters

TRANSMITTER

Operating frequency	13.75 GHz and 13.76 GHz (frequency agility)
Peak power	1000 W
Pulsewidth	1.67 μs
PRF	3528 Hz
Average power	12 W

ANTENNA

Size	10 m × 10 m
Polarization	linear
3-dB beamwidth	0.18° × 0.18° (at nadir)
Gain	60 dB
Sidelobe level	< −30 dB
Scan angle:	±5.7°

RECEIVER (LOGARITHMIC)

Input bandwidth	50 MHz
Noise figure	2.5 dB
Dynamic range	70 dB
Total system losses	8 dB
Data rate	200 kb/s
Total power consumption	150 W
Weight	285 kg

Resolution and Swath

Range resolution: 250 m
Horizontal resolution (at nadir):   1.6 km
Three swaths: one central, two laterals (one selected by adaptative pointing)
Angular position of the swath center: −11.3°, 0°, +11.3°
Swathwidth: 100 km for each of the 3 swaths

Vertical resolution: $\alpha = 0°$:   250 m, $\alpha = 5.7°$: 400 m
                       $\alpha = 11.3°$: 560 m, $\alpha = 17°$: 700 m
Spacing between FOV: 3 km (cross and along track)

Detection and Accuracy

Nonambiguous range: 40 km (±20 km from the surface)
Minimum measurable apparent reflectivity: 8 dBZ
Range of measurable rainfall rate (SNR > 3 dB): 0.3–60 mm/h
(at bottom of 5 km depth uniform rain layer)
SNR: 0.3 < R < 60 mm/h   ≥ 3 dB
       0.7 < R < 50 mm/h   ≥10 dB
       1.5 < R < 40 mm/h   ≥15 dB

Table 3.12 continued

Detection and Accuracy

Dwell time, integration time: 14 ms, 11 ms
Number of independent samples: 60 (with frequency agility)
Power measurement accuracy: <0.7 dB (for SNR > 10 dB)

cant increase in the antenna area. The antenna proposed will be deployed apart from the spacecraft. The design is an unfurlable paraboloid, using an offset array feed. Because the paraboloid is fixed and scanning will be accomplished by steering the primary illumination pattern across the reflector surface, some concern exists as to the extent of scanning without degradation of the main beam and increases in the magnitude of the sidelobe levels.

REFERENCES

[1]   Okamoto, K., S. Miyazaki, and T. Ishida, 1979: Remote sensing of precipitation by a satellite-borne microwave remote sensor. *Acta Astronautica,* **6**, 1043–1060.

[2]   Stepanenko, V.D., 1966: *Radar in Meteorology.* Gidrometeoizdat (Leningrad). (Partial translation available from Joint Publications Research Service, Washington, DC)

[3]   Atlas, D., 1982: Adaptively pointing spaceborne radar for precipitation measurements. *J. Appl. Meteor.,* **21**, 429–431.

[4]   Simpson, J., ed., 1988: TRMM—A satellite mission to measure tropical rainfall. Report of the Science Steering Group. NASA/GSFC, August, 94 pp.

[5]   Spielman, B.E., 1986: Defense Electronics: A balance of functionality, finance and form. *Microwave J.,* **29**, 26–39.

[6]   Bierman, H., 1987: Transistors stride to mm-wave performance. *Microwave J.,* **30**, 30–41.

[7]   Lhermitte, R., 1981: Satellite borne dual millimetric wave length radar. In Precipitation Measurements from Space: Workshop Report, Atlas. D., and O.W. Thiele, eds. NASA/GSFC, Greenbelt, MD, D-277–D-282.

[8]   Lhermitte, R., 1989: Satellite-borne millimeter wave Doppler radar. URSI Commission F, Open Symposium, La Londe-Les-Maures, France, September 11–15.

[9]   Nathanson, F.E., T.H. Slocumb, L. Brooks, R.K. Crane, and S.W. McCandless, 1988: Radar Sounder. Final Report, Contract Number F19628-87-C-0231, 133 pp.

[10] North, G.R., 1988: Tropical rain mapping radar (TRAMAR), Vol. 1: Investigation and Technical Plan, Data Plan, Calibration Plan, Management Plan. Texas A&M and JPL. July, 1988. 40 pp. plus appendices.

[11] Im, E., and F. Li, 1989: Tropical rain mapping radar on the Space Station. *IGARSS '89,* Vancouver, British Columbia, July 10–14, 1485–1490.

[12] Im, K., F.K. Li, W.J. Wilson, and D. Rosing, 1987: Conceptual design of a spaceborne radar for global rain mapping. *IGARSS '87,* Ann Arbor, MI.

[13] Li, F., K. Im, W.J. Wilson, and C. Elachi, 1988: On the design issues for a spaceborne rain mapping radar. *Tropical Rainfall Measurements,* J.S. Theon and N. Fugono, eds., A Deepak Publishing, Hampton, VA, 387–393.

[14] Goldhirsh, J., 1988: Analysis of algorithms for the retrieval of rain rate profiles from a spaceborne dual-wavelength radar. *IEEE Trans. Geosci. and Remote Sens.,* **GE-26**, 98–114.

[15] BEST: Tropical System Energy Budget, 1988. Centre National d'Etudes Spatiales, October, 58 pp.

[16] Marzoug, M., P. Amayenc, J. Testud, and N. Karouche, 1989: Conceptual design of the spaceborne rain radar of the B.E.S.T. project. *Preprints 24th Conf. on Radar Meteor.*, March 27–31, Amer. Meteor. Soc., Boston, MA, 597–600.

[17] Okamoto, K., T. Kozu, K. Nakamura, and T. Ihara, 1988: Tropical rainfall measuring mission radar. Tropical Rainfall Measurements, J.S. Theon and N. Fugono, eds., A Deepak Publishing, Hampton, VA, 213–219.

[18] Okamoto, K., J. Awaka, and T. Kozu, 1988: A feasibility study of the rain radar for the tropical rainfall measuring mission: 6. A case study of rain radar systems. *J. Comm. Research Lab.*, **35**, 183–208.

[19] Katzenstein, H., and H. Sullivan, 1960: A new principle for satellite-borne meteorological radar. *Proc. 8th Weather Radar Conf.*, Amer. Meteor. Soc., Boston, MA, 505–515.

[20] Skolnik, M.I., 1974: The Application of Satellite Radar for the Detection of Precipitation. NRL Rept. 2896, October, 100 pp.

[21] NASA-JPL, 1986: Shuttle Imaging Radar-C Science Plan. JPL Publication 86-29, September.

[22] Eckerman, J., 1975: Meteorological radar facility for the space shuttle. *IEEE National Telecomm. Conf.*, New Orleans, IEEE Publication 75 CH1015 CSCB, 37-6–37-17.

[23] Okamoto, K., 1983: Remote sensing of precipitation by weather radar system at space station. *Preprints 21 Conf. on Radar Meteor.*, Amer. Meteor. Soc., Boston, MA, 263–269.

[24] Thiele, O.W., ed., 1987: On requirements for a satellite mission to measure tropical rainfall. NASA Ref. Publ. #1183, 49 pp.

[25] Ihara, T., and K. Nakamura, 1988: A feasibility study of the rain radar for the tropical rainfall measuring mission: 4. A discussion of pulse-compression and adaptive scanning. *J. Comm. Research Lab.*, **35**, 149–161.

[26] Awaka, J., T. Kozu, and K. Okamoto, 1988: A feasibility study of the rain radar for the tropical rainfall measuring mission: 2. Determination of basic system parameters., *J. Comm. Research Lab.*, **35**, 111–133.

[27] Nakamura, K., and T. Ihara, 1988: A feasibility study of the rain radar for the tropical rainfall measuring mission: 3. Radar type and antenna. *J. Comm. Research Lab.*, **35**, 135–148.

[28] Elachi, C., 1987: *Spaceborne Radar Remote Sensing: Applications and Techniques.* IEEE Press, New York, 255 pp.

[29] Silver, S., ed., 1949: *Microwave Antenna Theory and Design.* MIT Radiation Lab. Series, **12**, McGraw-Hill, New York, 623 pp.

[30] Cheston, T.C., and J. Frank, 1970: *Array Antennas.* Ch. 11, M.I. Skolnik, ed., McGraw-Hill, New York.

[31] Rudge, A.W., 1975: Multiple-beam antennas: Offset reflectors with offset feeds. *IEEE Trans. Ant. and Propag.*, **AP-23**, 317–322.

[32] Radford, M.A., 1978: Electronically scanned antenna systems. *Proc. IEE*, **11R**, 1100–1112.

[33] Stark, L., R.W. Burns, and W.P. Clark, 1970: Phase Shifters for Arrays. Ch. 11 in *Radar Handbook*, Skolnik, M.I., ed., McGraw-Hill, New York.

[34] Cook, C.E., and M. Bernfeld, 1967: *Radar Signals: An Introduction of Theory and Application.* Academic Press, New York.

[35] Gray, R.W., and D.T. Farley, 1973: Theory of incoherent-scatter measurements using aom prouood pulses. *Radio Sci.*, **8**, 123–131.

[36] Raney, R.K., 1980: SAR response to partially coherent phenomena. *IEEE Trans. Ant. and Propag.*, **AP-28**, 777–787.

[37] Ulaby, F.T., R.K. Moore, and A.K. Fung, 1982: *Microwave Remote Sensing: Active and Passive.* Artech House, Norwood, MA, Vol. 2, 608 pp.

[38] Brunfeldt, D.R., and F.T. Ulaby, 1984: Active reflector for radar calibration. *IEEE Geosci. and Remote Sens.*, **GE-22**, 165–169.

[39] Moore, R.K., and M. Hemmat, 1988: Determination of the vertical pattern of the SIR-B antenna. *Int. J. Remote Sensing*, **9**, 839–847.

[40] Joss, J., R. Cavalli, and R.K. Crane, 1974: Good agreement between theory and experiment for attenuation data. *J. Rech. Atmos.*, **8**, 299–318.

[41] Goldhirsh, J., 1979: A review of the application of nonattenuating frequency radars for estimating rain attenuation and space-diversity performance. *IEEE Trans. Geosci. and Remote Sens.*, **Ge-17**, 218–239.

[42] Kozu, T., K. Nakamura, J. Awaka, and M. Takeuchi, 1987: Development of Ku-band FM-CW/pulse compression radar for rain observation on a slant-path. *J. Radio Res. Lab.*, **34**, 95–113.

[43] Hodge, D.B., and G.L. Austin, 1977: The comparison between radar- and radiometer-derived rain attenuation for earth-space links. *Radio Sci.*, **12**, 733–740.

[44] Atlas, D., and R.K. Moore, 1987: The measurement of precipitation with synthetic aperture radar. *J. Atmos. Ocean. Technol.*, **4**, 368–376.

[45] Atlas, D., C. Elachi, and W.E. Brown, Jr., 1977: Precipitation mapping with an airborne synthetic aperture imaging radar. *J. Geophys. Res.*, **82**, 3445–3451.

[46] Metcalf, J.I., and W.A. Holm, 1979: Meteorological applications of synthetic aperture radar. Final Rep. Proj. A-2101 Eng. Experiment Station, Georgia Inst. of Tech., 33 pp.

[47] NASA, 1984: Earth Observing System—Science and Mission Requirements Working Group Report. Vol. 1. Tech. Memo. 86129, 51 pp. Goddard Space Flight Center, Greenbelt, MD, 20771.

[48] Bucknam, J., R.P. Dooley, A. Fredriksen, F.E. Nathanson, 1975: Synthetic Azimuth Processing Mode—Concept Analysis, Technical Note 3, Rev. 1, Technology Service Co., NASA/GSFC Contract NAS5-20058, March.

[49] Bucknam, J., R.P. Dooley, F.E. Nathanson, 1975: Shuttle Meteorological Radar Study, Final Report, Technology Service Co., NASA/GSFC Contract NAS5-20058, April.

[50] Lhermitte, R., 1988: Cloud and precipitation sensing at 94 GHz. *IEEE Trans. Geosci. and Remote Sens.*, **GE-26**, 207–216.

[51] Nathanson, F.E., T.H. Slocumb, S.W. McCandless, and R.K. Crane, 1989: A space based radar to measure clouds and rain. *IGARSS '89*, Vancouver, Canada, July 10–14, 1984.

[52] Goldhirsh, J., and E.J. Walsh, 1982: Rain measurements from space using a modified Seasat-Type altimeter. *IEEE Trans. Antennas and Propag.*, **AP-30**, 726–733.

[53] Moore, R.K., and S.S. Xie, 1988: Radar rainfall measurement system using SAR and conical-scan methods. *Tropical Rainfall Measurements*, J.S. Theon and N. Fugono, eds., A Deepak Publishing, Hampton, VA, 243–253.

[54] Simpson, J., R.F. Adler, and G.R. North, 1988: A proposed tropical rainfall measuring mission. *Bull. Amer. Meteor. Soc.*, **69**, 278–295.

[55] Okamoto, K., 1988: A feasibility study of the rain radar for the tropical rainfall measuring mission: 1. Introduction. *J. Comm. Research Lab.*, **35**, 109–110.

[56] Manabe, T., and T. Ihara, 1988: A feasibility study of the rain radar for the tropical rainfall measuring mission: 5. Effects of surface clutter on rain measurements from satellite. *J. Comm. Research Lab.*, **35**, 163–181.

Chapter 4
Meteorological-Radar Relationships

For a conventional radar (single polarization, non-doppler), the coupling between the mean radar return power and meteorological parameters arises through two factors: the effective radar reflectivity factor, Z_e, and the two-way attenuation, $\exp(-0.2 \ln 10 \int k \, ds)$. In this chapter, we begin with a brief account of absorption by atmospheric gases, and follow this with a discussion of the backscattering and attenuation properties of cloud and precipitation.

4.1 ABSORPTION FROM ATMOSPHERIC GASES

For frequencies below 350 GHz, absorption from gases other than water vapor and molecular oxygen is insignificant. Therefore, the specific attenuation, k_g, in dB/km from atmospheric gases (due to absorption) can be written as:

$$k_g = k(H_2O) + k(O_2) \tag{4.1}$$

For water vapor, the centers of the first two absorption lines occur at $f = 22.235$ GHz and $f = 183.31$ GHz. For oxygen, a large number of absorption lines exist in the region from 50 to 70 GHz, with an additional line at 118.75 GHz. Pressure broadening in the lower atmosphere causes the individual lines to merge into a continuous distribution with a maximum at approximately 60 GHz [1].

For $f < 100$ GHz, $k(H_2O)$ can be expressed as [1–2]:

$$k(H_2O) = 2f^2\rho(300/T)^{3/2}\gamma$$

$$\times \{(300/T)e^{-644/T}[(494.4 - f^2)^2 + 4f^2\gamma^2]^{-1} + 1.2 \times 10^{-6}\} \text{ dB/km} \tag{4.2}$$

where ρ is the water vapor density (g m^{-3}), T is the temperature in kelvins, and the line width parameter, γ, is given by

$$\gamma = 2.85(P/1013)(300/T)^{0.626}(1 + 0.018\rho T/P) \text{ (GHz)} \tag{4.3}$$

where P is the atmospheric pressure in millibars and where both f and γ are in units of GHz. Expressions for $f > 100$ GHz can be found in references [1] and [2].

For oxygen, the following expression is valid for frequencies less than 45 GHz [2–3]

$$k(O_2) = 1.1 \times 10^{-2}f^2(P/1013)(300/T)^2$$

$$\times \gamma\{[(f - f_0)^2 + \gamma^2]^{-1} + (f^2 + \gamma^2)^{-1}\} \text{ (dB/km)} \tag{4.4}$$

where $f_0 = 60$ GHz and the linewidth parameter is

$$\gamma = \gamma_0(P/1013)(300/T)^{0.85} \text{ (GHz)} \tag{4.5}$$

with

$$\gamma_0 = \begin{cases} 0.59, & P \geq 333 \text{ mbar} \\ 0.59[1 + 3.1 \times 10^{-3}(333 - P)], & 25 \leq P \leq 333 \text{ mbar} \\ 1.18, & P \leq 25 \text{ mbar} \end{cases} \tag{4.6}$$

For precipitation and cloud sensing, the quantity of greatest interest is the total absorption by gases to the observational range. This quantity is shown in Figure 4.1 [4] for $f < 100$ GHz at various starting heights above the surface. If the radar range is located at an altitude z above the surface at an off-nadir angle θ, then the two-way path attenuation (dB) from atmospheric absorption can be written as:

$$A_g = \exp(-0.2 \ln 10 \int_0^r k_g \, ds) = \exp[-0.2 \ln 10 \sec \theta \int_z^H k_g(z) \, dz] \tag{4.7}$$

where H is the satellite altitude (effectively at infinity). Values of $\int_z^\infty k_g(z) \, dz$ can be read from the curves by specifying the frequency and "starting altitude," z. More generally, equations for the total vertical absorption have been expressed as functions of the mean absolute humidity and temperature by means of a regression of the form [4]:

$$\int_0^\infty k_g(z) \, dz = a + b\rho - cT \tag{4.8}$$

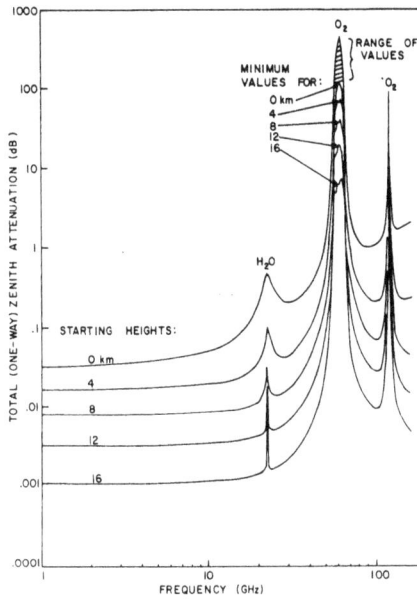

Figure 4.1 Theoretical one-way zenith attenuation from specified height to top of atmosphere for a moderately humid atmosphere ($\rho = 7.5$ g/m^{-3} at the surface) (from Crane and Blood [4]).

where T is the mean local surface temperature (°C) and ρ is the mean absolute local surface humidity (g m^{-3}). Regression coefficients a, b, and c for selected frequencies are shown in Table 4.1 [4]. Global maps of ρ and T are given by Bean and Dutton [5].

Table 4.1 Coefficients for Computing Total Zenith Attenuation [4]
$$\tau_{90} = a + b\rho_0 - cT_0 \text{ (dB)}$$

Frequency	Coefficients		
F(GHz)	$a(F)$	$b(F)$	$c(F)$
1	3.3446E-02	2.7551E-00	1.1189E-04
4	3.9669E-02	2.7599E-04	1.7620E-04
6	4.0448E-02	6.5086E-04	1.9645E-04
12	4.3596E-02	3.1768E-03	3.1470E-04
15	4.6138E-02	6.3384E-03	4.5527E-04
16	4.7195E-02	8.2112E-03	5.3568E-04
20	5.6047E-02	3.4557E-02	1.5508E-03

Table 4.1 continued

Frequency F(GHz)	Coefficients		
	$a(F)$	$b(F)$	$c(F)$
22	7.5989E-02	7.8251E-02	3.0978E-03
24	6.9102E-02	5.9116E-02	2.4950E-03
30	8.5021E-02	2.3728E-02	1.3300E-03
35	1.2487E-01	2.3681E-02	1.4860E-03
41	2.3683E-01	2.8402E-02	2.1127E-03
45	4.2567E-01	3.2766E-02	2.9945E-03
50	1.2671E 00	3.9155E-02	5.7239E-03
55	2.4535E 01	4.8991E-02	−1.2125E-03
70	2.1403E 00	7.3246E-02	1.0436E-02
80	7.0496E-01	9.5860E-02	5.8635E-03
90	4.5760E-01	1.2185E-01	5.7369E-03
94	4.1668E-01	1.3320E-01	5.9439E-03
110	4.3053E-01	1.8465E-01	7.8499E-03
115	8.9351E-01	2.0292E-01	1.1297E-02
120	5.3532E 00	2.2125E-01	3.6311E-02
140	3.6788E-01	3.1894E-01	1.1941E-02
160	4.1446E-01	5.0635E-01	1.9078E-02
180	2.8087E 00	5.0360E 00	1.9198E-01
200	5.6172E-01	8.9655E-01	3.3943E-02
220	5.4358E-01	7.7720E-01	2.7580E-02
240	6.0124E-01	8.7887E-01	3.0693E-02
280	7.5941E-01	1.2220E 00	4.2753E-02

Atmospheric gas attenuation is only one component of the path attenuation. The total attenuation (dB) from the radar to the range r can be expressed as

$$A(r) = \exp\left[-0.2 \ln 10 \int_0^r (k_g + k_c + k_p) \, ds\right] \tag{4.9}$$

where k_c and k_p are the specific attenuations in dB/km for clouds and precipitation, respectively. Figure 4.2 shows the contributions from k_g, k_c, and k_p as a function of frequency [4]. While for Rayleigh scattering, k_c is determined by the cloud liquid water, temperature and frequency, k_p depends not only on these parameters but also on the distribution of drop sizes: for Figure 4.2, a Laws and Parsons [6] drop size distribution has been assumed. This figure is useful in identifying the range of frequencies for which a cloud or precipitation radar can function without excessive attenuation from atmospheric gases; in particular, the atmospheric window in the vicinity of 94 GHz is considered one of the primary candidates for a cloud radar (see Section 3.7).

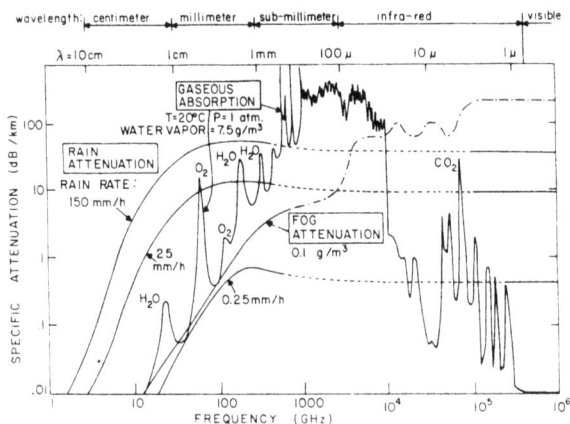

Figure 4.2 Specific attenuations due to gaseous constituents and precipitation for transmission within the atmosphere (from Crane and Blood [4]).

4.2 DEFINITIONS OF METEOROLOGICAL AND RADAR PARAMETERS

A list of commonly used quantities in radar meteorology is given in Table 4.2, along with their definitions and units. Although the units are somewhat typical, there is no standard set. Some caution, therefore, is needed in comparing various results in the literature. We note several other points:

(1) Some ambiguity exists in the definitions of M and R for the case of nonliquid hydrometeors. The drop size distribution (DSD) is usually specified as a function of the diameter of the melted particle so that $\rho = \rho_w$ where ρ_w is the mass density of water.

(2) The quantities η and Z_e are more general than Z, which applies only to Rayleigh scattering of spherical water drops.

(3) The $|K_w|^{-2}$ dependence in the definition of Z_e, when substituted into the radar equation, cancels an identical factor in the numerator. Thus, the only quantities in the radar equation that depend on the refractive index are the backscattering and extinction cross sections of the hydrometeors.

(4) Various conventions have been used in the literature for the definitions of Z and Z_e for nonliquid hydrometeors [7].

(5) Generalizations of the radar parameters are required for bistatic scattering (see Section 2.6) and in cases where an account of the polarization properties of the scatterers must be made (see Section 2.7). In particular, when the particle shapes

Table 4.2

Quantity	Definition	Units
σ_b	Backscattering Cross Section	cm^2
σ_a	Absorption Cross Section	cm^2
σ_s	Scattering Cross Section	cm^2
σ_t	Total (Extinction) Cross Section	cm^2
	$\sigma_a + \sigma_s$	
ρ	Mass Density of Hydrometeors	g cm^{-3}
λ	Wavelength	cm
m	Complex Index of Refraction [for exp($i\omega t$) convention, $m_I > 0$]	
$(m_R - im_I)$		
K	Dielectric Factor	
	$(m^2 - 1)/(m^2 + 2)$	
D	Drop Diameter	cm
$N(D)$	Drop Size Distribution (*DSD*)	m^{-3} cm^{-1}
$v(D)$	Distribution of Velocities	m s^{-1}
η	Radar Reflectivity	cm^{-1}
	$10^{-6} \int \sigma_b(D)N(D)\,dD$	
Z_e	Equivalent (Effective) Radar Reflectivity Factor	mm^6 m^{-3}
	$\dfrac{10^6 \lambda^4}{\pi^5 |K_w|^2} \int \sigma_b(D)N(D)\,dD$	
Z	Radar Reflectivity Factor	mm^6 m^{-3}
	$10^6 \int D^6 N(D)\,dD$	
k	Specific Attenuation	dB km^{-1}
	$0.434 \int \sigma_t(D)N(D)\,dD$	
M	Equivalent Liquid Water Content	g m^{-3}
	$\dfrac{\pi\rho}{6} \int D^3 N(D)\,dD$	
R	Rain Rate	mm h^{-1}
	$0.6\pi \int v(D)N(D)D^3\,dD$	

deviate from sphericity, additional integrations are needed over the distribution of orientations to define the radar parameters in Table 4.2. In certain cases of Rayleigh scattering, however, the scattering characteristics can be approximated by spheres of equal mass [7–8].

(6) For hydrometeors in the melting layer, which are often modeled as an inner core of snow surrounded by a water shell, the distribution of sizes is determined by the snow density, the fractional mass of the particle that has melted and by the outer, D_0, (or inner, D_i) diameter [9–10]. For these models, the size distribution is often expressed as a function of the outer diameter rather than the melted diameter. Assuming that a one-to-one correspondence exists between D_0 and D_i at a fixed altitude h so that $D_i = f(D_0, h)$, then in terms of the outer diameter, M and R become

$$M = \frac{\pi}{6} \int \rho(D_0) D_0^3 N(D_0) \, dD_0 \tag{4.10}$$

$$R = (0.6\pi/\rho_w) \int v(D_0) D_0^3 N(D_0)\rho(D_0) \, dD_0 \tag{4.11}$$

with

$$\rho(D_0) = \left| \rho_s D_i^3 + \rho_w (D_0^3 - D_i^3) \right| / D_0^3 \tag{4.12}$$

where ρ_s is the density of snow, which ranges from about 0.05 to 0.2 gm cm^{-3}. For graupel, the mass density can vary from less than 1 gm cm^{-3} to almost half that of ice.

4.3 DROP SIZE DISTRIBUTIONS (DSD)

We can see from Table 4.2 that the drop size distribution, $N(D)$, furnishes the primary link between the radar quantity and the meteorological quantity. One of the most general forms of the size distribution is given by the modified gamma distribution [11]:

$$N(D) = N_0 D^m \exp(-\Lambda D^q) \tag{4.13}$$

Using the form given by (4.13), Deirmendjian [11] has characterized various cloud types by a selection of the four free parameters in the distribution [14–15].

Although a complete specification of the DSD requires a knowledge of the maximum drop diameter, D_{max}, if the median drop diameter is much less than D_{max}, then the error incurred by extending the integration to infinity is small [12–13]. We note, however, that the errors will be larger for such quantities as the reflectivity, which involve higher order moments of the DSD. Assuming that the integration can be extended to infinity, the νth moment of the distribution is

$$M(m,q,\nu) = \int_0^\infty D^\nu N(D) \; dD = N_0 \Gamma(\alpha)/|q|\Lambda^\alpha \tag{4.14}$$

where $\alpha = (m + 1 + \nu)/q$, and where Γ is the complete gamma function. The mean and standard deviation, therefore, are given, respectively, by $M(m,q,\nu = 1)$ and

$$\sigma = [M(m,q,\nu = 2) - M^2(m,q,\nu = 1)]^{0.5} \tag{4.15}$$

while the total number of drops, N_T, per unit volume is simply $M(m,q,\nu = 0)$. Other relevant statistics of the distribution are the mode diameter (the value of D for which the distribution is maximum) given by $D = (m/q\Lambda)^{1/q}$ and the median mass diameter, D_0, defined as the diameter for which half the total mass is contained in drops of diameter less than D_0:

$$2 \int_0^{D_0} D^3 N(D) \; dD = M(m,q,\nu = 3) \tag{4.16}$$

For $q = 1$, (4.13) reduces to the standard gamma distribution, which has been used to characterize both rain [13,16] and hail [17]:

$$N(D) = N_0 D^m \exp(-\Lambda D) \tag{4.17}$$

where m and Λ are related to the median drop diameter by the approximate expression [13]:

$$D_0 = (3.67 + m)/\Lambda \tag{4.18}$$

For rain, the parameter m has been found to vary between -2 and 8, where negative values of m correspond to concave upward distributions, which sometimes can be used to characterize orographic rain. We note, however, that for $m \leq -1$, the number density is undefined (i.e., the integral of $N(D)$ is unbounded). In the majority of cases, the value of m tends to be positive [13,18]. Recent dual-polarization measurements, for example, have shown improved correlations between the differential reflectivity (the difference, in dB, of the copolarized returns), and either D_0 or the rain rate if values of m between 2 and 4 are chosen [19].

Still the most commonly used *DSD* is the exponential, which follows from (4.13) with $m = 0$ and $q = 1$

$$N(D) = N_0 \exp(-\Lambda D) \tag{4.19}$$

where $D_0 = 3.67/\Lambda$. This distribution has been used extensively to describe rain, snow, hail and cloud. For nonliquid hydrometeors, the usual convention is to interpret D as the diameter of a water sphere of mass equal to that of the frozen particle. In many instances, the parameters N_0 and Λ of the distribution have been expressed as functions of the rain rate, R. For widespread stratiform [21], thunderstorm, and drizzle types of rain [22], values of N_0, and Λ are, respectively, [8 × 10^4, 41 $R^{-0.21}$], [1.4 × 10^4, 30 $R^{-0.21}$], and [3 × 10^5, 57 $R^{-0.21}$], while for snow the parameters have been given as [2.5 × $10^4 R^{-0.94}$, and 22.9 $R^{-0.45}$] [12]. In all cases, N_0 is in m^{-3}/cm^{-1}, Λ is in cm^{-1} and R in mm/h^{-1}. Olsen et al. [23] have pointed out that these values, when substituted into the expression for rain rate, do not always satisfy the resultant integral equation for R:

$$R = 0.6\pi N_0(R) \int D^3 v(D) e^{-\Lambda(R)D} \, dD$$

Using, for example, the relationship $v(D) = aD^b$ with $a = 17.67$, and $b = 0.67$ (see Section 4.4), an integration of this equation over all D gives

$$R = 0.6\pi a N_0(R)\Gamma(4 + b)[\Lambda(R)]^{-(4+b)}$$

To satisfy this in the case of the Marshall-Palmer distribution [21], holding N_0 fixed at 8 × 10^4, $\Lambda(R)$ must be modified from 41 $R^{-0.21}$ to 42.3 $R^{-0.214}$. If, on the other hand, $\Lambda(R)$ is held constant, as suggested by Olsen et al. [23], then N_0 must be changed to 6.91 × $10^4 R^{0.02}$.

Although the log-normal DSD has not been as widely used as the various forms of the gamma distribution, the distribution has found applications to both rain and cloud [24–26]. This distribution can be written in the form:

$$N(D) = (N_T/\sqrt{2\pi} \, \sigma D) \exp[-0.5((\ln D - m)/\sigma)^2] \qquad (4.20)$$

The vth moment is

$$M(\sigma,m,v) = N_T \exp[mv + (\sigma v)^2/2] \qquad (4.21)$$

A detailed description of the distribution and the fitting procedure to measured distributions can be found in references [24–26].

4.4 VELOCITY DISTRIBUTIONS

Because the rainfall rate is related to the downward flux of water, it is dependent upon the velocity distribution of drops. For the majority of spaceborne weather radars that have been proposed, the velocity will not be a measurable quantity.

Therefore, the terminal fall speeds of the hydrometeors must be estimated. This is the case even with doppler radar where the calculation of the updraft requires a knowledge of the mean terminal speed of the scatterers (see Section 2.8). Doppler measurements within the bright-band, moreover, may provide information on coalescence and drop shattering if prior relationships can be established between velocity and parameters such as size, shape, and thermodynamic phase.

For cloud droplets less than 40 μm, which includes the majority of the population, the validity of Stokes's Law implies that the fall speed of the droplet is proportional to the square of the radius:

$$v = 1.19 \times 10^6 r^2 \text{ (cm s}^{-1}) \tag{4.22}$$

In this and the following formulas, the radius, r, or diameter, D, is expressed in cm.

For radii between 40 μm and 0.6 mm, which spans the range from large cloud droplets to very small precipitation drops, Rogers [15] gives the approximation:

$$v = 8 \times 10^3 r \text{ (cm s}^{-1}) \tag{4.23}$$

For small raindrops, the following approximation can be used [27–28]:

$$v = 2.01 \times 10^3 (\rho_0/\rho(z))^{0.5} r^{0.5} \text{ (cm s}^{-1}) \tag{4.24}$$

where $\rho(z)$ is the air density at height z, and ρ_0 is the reference value equal to 1.2 $\times$ 10^{-3}, which corresponds to the density of dry air at temperature 20°C and pressure of 1013 mbar.

Due to the great variation in shape and orientation, the velocity-size relationships for single ice crystals, snow aggregates and graupel are much harder to quantify. Figure 4.3 show relationships between velocity and size for various types of dry snow crystals, partially melted crystals and graupel (opaque clumps of frozen droplets usually in a spherical or conical shape). For ice crystals, the velocity can be approximated by:

$$v(D) = aD^{0.31} \text{ (m s}^{-1}) \tag{4.25}$$

where $a = 1.6$ for dendrites, $a = 2.34$ for crystals in the form of columns and plates, and $a = 2$ for aggregate snowflakes [29]. In this formula, D is the melted diameter in cm. An approximation to the fall speed for graupel is [15]:

$$v(D) = 3.43D^{0.6} \text{ (m s}^{-1}) \tag{4.26}$$

where D is the diameter (cm) of the sphere which circumscribes the particle.

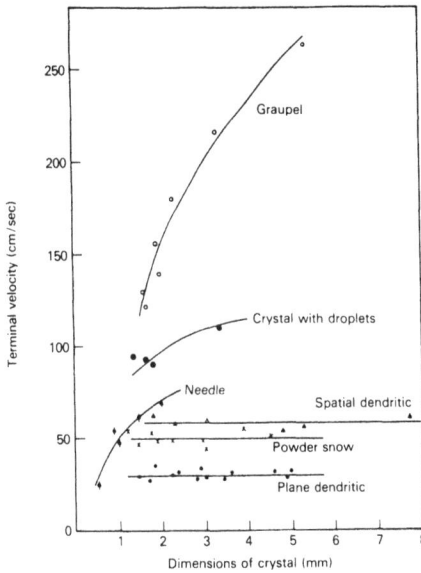

Figure 4.3 Measured terminal velocity of differential types of ice crystals (from Fletcher [49], after Nakaya and Terada).

Expressions for the velocity-size relationships have also been developed for melting snow aggregates within the bright-band [10,30,31].

For raindrops, the data of Gunn and Kinzer [32] are widely used and can be modified to account for variations in height by multiplying by the factor $(\rho_0/\rho(z))^{0.5}$ as described above. Plots of the raindrop velocity *versus* drop diameter at several heights are shown in Figure 4.4 [33]. It is useful to approximate these data by expressions that can be integrated analytically when multiplied by drop size distributions such as the exponential or gamma. Three approximations to the velocity distributions of raindrops are

$$v(D) = 14.2D^{0.5} \qquad \text{(Spilhaus [34])} \qquad (4.27)$$

$$v(D) = 17.67D^{0.67} \qquad \text{(Atlas and Ulbrich [35])} \qquad (4.28)$$

$$v(D) = 9.25 - 9.25 \exp[-(6.8D^2 + 4.88D)] \qquad \text{(Lhermitte, [36]) } (4.29)$$

In all cases, $v(D)$ is in m s^{-1}, while D is in cm. We note that for hail, an expression similar to (4.27) is commonly used [37]:

$$v(D) = 16.2D^{0.5} \qquad (4.30)$$

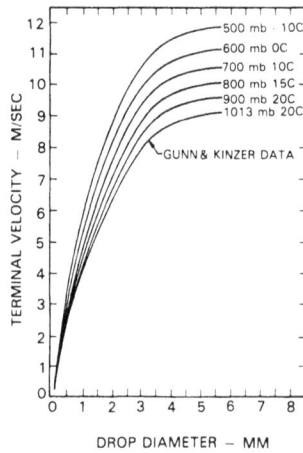

Figure 4.4 Terminal velocities of raindrops *versus* drop diameter for several temperatures and pressures (from Foote and duToit, [33]).

The data of Gunn and Kinzer are shown in Table 4.3 along with the results obtained from (4.27)–(4.29). The best approximation to the measured data is given by (4.29); (4.28), however, is a simpler expression and provides a fairly good fit to the measurements in the diameter range between 0.6 and 3.4 mm. It has been pointed out that the error in the approximation given by (4.27) arises from the fact that while the diameter dependence given by $D^{0.5}$ is valid for spherical drops, as the drop volume increases, the actual drop shape takes on an oblate spheroidal shape which, with further growth, exhibits flattening at the base [38].

Table 4.3 Gunn and Kinzer Data of Drop Velocity *versus* Diameter and Comparisons of Three Approximations

	Velocity (m/s)			
Diameter (cm)	Gunn-Kinzer	(4.27)	(4.28)	(4.29)
0.01	0.27	1.42	0.81	0.45
0.02	0.72	2.01	1.29	0.88
0.03	1.17	2.46	1.69	1.31
0.04	1.62	2.84	2.04	1.72
0.05	2.06	3.18	2.37	2.12
0.06	2.47	3.48	2.68	2.51
0.07	2.87	3.76	2.97	2.89
0.08	3.27	4.02	3.25	3.26
0.09	3.67	4.26	3.52	3.61
0.10	4.03	4.49	3.78	3.95

Table 4.3 continued

Diameter (cm)	Velocity (m/s)			
	Gunn-Kinzer	(4.27)	(4.28)	(4.29)
0.12	4.64	4.92	4.27	4.58
0.14	5.17	5.31	4.73	5.16
0.16	5.65	5.68	5.18	5.69
0.18	6.09	6.02	5.60	6.17
0.20	6.49	6.35	6.01	6.59
0.22	6.90	6.66	6.41	6.98
0.24	7.27	6.96	6.79	7.31
0.26	7.57	7.24	7.17	7.61
0.28	7.82	7.51	7.53	7.87
0.30	8.06	7.78	7.89	8.09
0.32	8.26	8.03	8.24	8.28
0.34	8.44	8.28	8.58	8.45
0.36	8.60	8.52	8.91	8.59
0.38	8.72	8.75	9.24	8.71
0.40	8.83	8.98	9.56	8.81
0.42	8.92	9.20	9.88	8.89
0.44	8.98	9.42	10.19	8.96
0.46	9.03	9.63	10.50	9.02
0.48	9.07	9.84	10.81	9.06
0.50	9.09	10.04	11.11	9.10
0.52	9.12	10.24	11.40	9.13
0.54	9.14	10.43	11.69	9.16
0.56	9.16	10.63	11.98	9.18
0.58	9.17	10.81	12.27	9.19

4.5 REFRACTIVE INDICES

The scattering and attenuation properties of the hydrometeors depend on the complex index of refraction, m, which in turn is a function of the temperature and phase state of the scatterer. In particular, the radar reflectivity and specific attenuation are proportional to $|K|^2$ and $\text{Im}\,(-K)$, respectively, where K is defined in terms of m by:

$$K = (m^2 - 1)/(m^2 + 2) \tag{4.31}$$

In some of the literature, the relative dielectric constant, e, the ratio of the permittivity of the hydrometeor to that of free space, is used rather than the index of refraction where the two quantities are related by $e = m^2$.

Figure 4.5 shows the real (m_R) and imaginary (m_I) part of the complex indices of refraction ($m = m_R - im_I$) for both water (top) and ice (bottom) in the

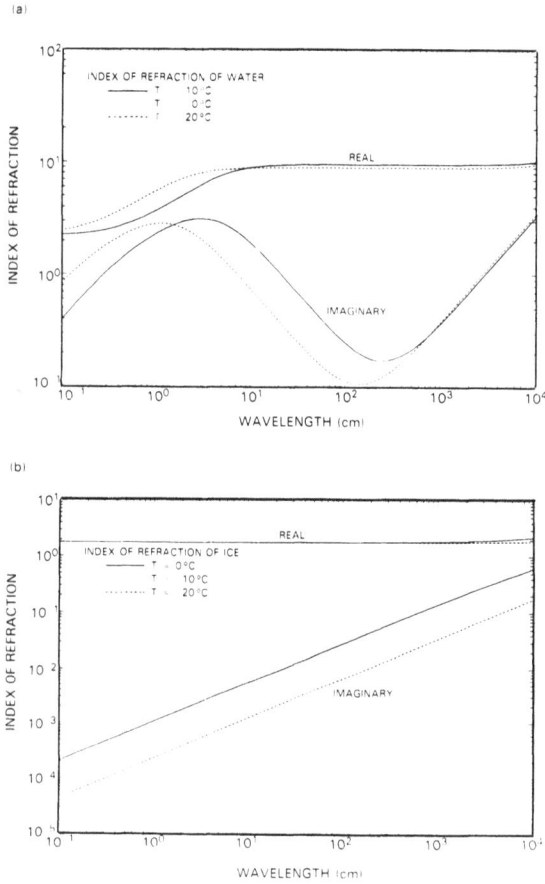

Figure 4.5 Complex indices of refraction for (a) water and (b) ice (after Ray [39]).

millimeter and microwave regions based on the model of Ray [39]. Because of a lack of measured data, uncertainty exists over the values for the imaginary part of ice. For most practical situations, however, the discrepancies among the various models are not important, and the imaginary part can often be neglected altogether. Much more important to attenuation measurements is the way the imaginary part of water reacts to changes in temperature and wavelength. For example, at X-band ($\lambda = 3$ cm), the imaginary part decreases significantly as the temperature increases from 0 to 20°C, whereas at Ka-Band ($\lambda = 0.86$ cm), the imaginary part is fairly insensitive to temperature changes. Results from the model are shown in Table 4.4 at selected wavelengths for temperatures ranging from −20°C to 30°C. Although $|K|^2$ is often taken to be 0.93, the error in this approximation will generally increase for shorter wavelengths and colder drops.

Table 4.4(a) Refractive Indices and Related Quantities for Water

λ (cm)	T (°C)	$m = (m_R - im_I)$ m_R	m_I	$\|K\|^2$	$Im(-K)$
10.000	-20.000	0.8556E+01	0.2562E+01	0.9393E+00	0.1983E-01
10.000	-10.000	0.8936E+01	0.1915E+01	0.9368E+00	0.1410E-01
10.000	0.000	0.9035E+01	0.1394E+01	0.9342E+00	0.1034E-01
10.000	10.000	0.8983E+01	0.1014E+01	0.9313E+00	0.7809E-02
10.000	20.000	0.8858E+01	0.7471E+00	0.9283E+00	0.6053E-02
10.000	30.000	0.8699E+01	0.5601E+00	0.9251E+00	0.4809E-02
3.000	-20.000	0.5324E+01	0.3158E+01	0.9268E+00	0.6527E-01
3.000	-10.000	0.6270E+01	0.3149E+01	0.9295E+00	0.4655E-01
3.000	0.000	0.7089E+01	0.2907E+01	0.9298E+00	0.3418E-01
3.000	10.000	0.7683E+01	0.2506E+01	0.9287E+00	0.2578E-01
3.000	20.000	0.8033E+01	0.2058E+01	0.9267E+00	0.1994E-01
3.000	30.000	0.8185E+01	0.1648E+01	0.9241E+00	0.1580E-01
1.870	-20.000	0.4209E+01	0.2737E+01	0.9054E+00	0.1016E+00
1.870	-10.000	0.5007E+01	0.2960E+01	0.9169E+00	0.7325E-01
1.870	0.000	0.5815E+01	0.3023E+01	0.9224E+00	0.5415E-01
1.870	10.000	0.6553E+01	0.2897E+01	0.9243E+00	0.4101E-01
1.870	20.000	0.7142E+01	0.2616E+01	0.9240E+00	0.3179E-01
1.870	30.000	0.7540E+01	0.2254E+01	0.9225E+00	0.2520E-01
1.240	-20.000	0.3480E+01	0.2271E+01	0.8646E+00	0.1437E+00
1.240	-10.000	0.4112E+01	0.2575E+01	0.8921E+00	0.1060E+00
1.240	0.000	0.4788E+01	0.2801E+01	0.9076E+00	0.7958E-01
1.240	10.000	0.5479E+01	0.2901E+01	0.9154E+00	0.6083E-01
1.240	20.000	0.6131E+01	0.2851E+01	0.9187E+00	0.4742E-01
1.240	30.000	0.6681E+01	0.2662E+01	0.9193E+00	0.3772E-01
0.857	-20.000	0.2995E+01	0.1848E+01	0.7977E+00	0.1849E+00
0.857	-10.000	0.3494E+01	0.2161E+01	0.8489E+00	0.1420E+00
0.857	0.000	0.4032E+01	0.2450E+01	0.8806E+00	0.1097E+00
0.857	10.000	0.4608E+01	0.2674E+01	0.8989E+00	0.8539E-01
0.857	20.000	0.5200E+01	0.2797E+01	0.9088E+00	0.6733E-01
0.857	30.000	0.5770E+01	0.2799E+01	0.9133E+00	0.5391E-01
0.319	-20.000	0.2297E+01	0.9019E+00	0.4948E+00	0.2109E+00
0.319	-10.000	0.2557E+01	0.1124E+01	0.5971E+00	0.2006E+00
0.319	0.000	0.2810E+01	0.1379E+01	0.6856E+00	0.1876E+00
0.319	10.000	0.3073E+01	0.1652E+01	0.7583E+00	0.1702E+00
0.319	20.000	0.3359E+01	0.1929E+01	0.8135E+00	0.1499E+00
0.319	30.000	0.3673E+01	0.2191E+01	0.8523E+00	0.1293E+00
0.214	-20.000	0.2185E+01	0.6351E+00	0.3949E+00	0.1724E+00
0.214	-10.000	0.2393E+01	0.8069E+00	0.4852E+00	0.1783E+00
0.214	0.000	0.2574E+01	0.1014E+01	0.5695E+00	0.1844E+00
0.214	10.000	0.2747E+01	0.1253E+01	0.6499E+00	0.1860E+00
0.214	20.000	0.2929E+01	0.1513E+01	0.7233E+00	0.1806E+00
0.214	30.000	0.3129E+01	0.1783E+01	0.7852E+00	0.1685E+00

For dry snow (i.e., mixtures of ice and air), the Debye formula can be used [20,40]. Letting e_m be the dielectric constant of the medium (matrix), e_{in} that of the inclusions or embedded particulates which fill a fraction f of the total particle volume, and e the dielectric constant of the mixture, then [40]

$$\frac{e-1}{e+2} = \frac{e_{in}-1}{e_{in}+2} f + \frac{e_m-1}{e_m+2}(1-f) \tag{4.32}$$

Table 4.4(b) Refractive Indices and Related Quantities for Ice

$$m = (m_R - im_I)$$

λ (cm)	T (°C)	m_R	m_I	$|K|^2$	$Im(-K)$
10.000	-20.000	0.1780E+01	0.4409E-04	0.1760E+00	0.1763E-04
10.000	-10.000	0.1780E+01	0.6089E-04	0.1760E+00	0.2434E-04
10.000	0.000	0.1780E+01	0.2065E-03	0.1760E+00	0.8257E-04
3.000	-20.000	0.1780E+01	0.1844E-04	0.1760E+00	0.7374E-05
3.000	-10.000	0.1780E+01	0.2495E-04	0.1760E+00	0.9976E-05
3.000	0.000	0.1780E+01	0.8764E-04	0.1760E+00	0.3504E-04
1.870	-20.000	0.1780E+01	0.1310E-04	0.1760E+00	0.5237E-05
1.870	-10.000	0.1780E+01	0.1758E-04	0.1760E+00	0.7028E-05
1.870	0.000	0.1780E+01	0.6259E-04	0.1760E+00	0.2503E-04
1.240	-20.000	0.1780E+01	0.9727E-05	0.1760E+00	0.3889E-05
1.240	-10.000	0.1780E+01	0.1296E-04	0.1760E+00	0.5184E-05
1.240	0.000	0.1780E+01	0.4672E-04	0.1760E+00	0.1868E-04
0.857	-20.000	0.1780E+01	0.7445E-05	0.1760E+00	0.2977E-05
0.857	-10.000	0.1780E+01	0.9860E-05	0.1760E+00	0.3942E-05
0.857	0.000	0.1780E+01	0.3591E-04	0.1760E+00	0.1436E-04
0.319	-20.000	0.1780E+01	0.3640E-05	0.1760E+00	0.1455E-05
0.319	-10.000	0.1780E+01	0.4741E-05	0.1760E+00	0.1896E-05
0.319	0.000	0.1780E+01	0.1777E-04	0.1760E+00	0.7105E-05
0.214	-20.000	0.1780E+01	0.2726E-05	0.1760E+00	0.1090E-05
0.214	-10.000	0.1780E+01	0.3527E-05	0.1760E+00	0.1410E-05
0.214	0.000	0.1780E+01	0.1337E-04	0.1760E+00	0.5347E-05

Note that the formula is symmetrical with respect to the roles of the inclusions and the medium. If the medium is air with the ice particulates as the inclusions, the dielectric constant for air is approximately 1, so the last term in (4.32) can usually be neglected. Recognizing further that the fractional ice volume can be written as $(M_i/\rho_i)/(M/\rho)$, where M and ρ are the mass and mass density, respectively, and where the subscripted quantities denote ice and the unsubscripted quantities the air-ice mixture, then (4.32) can be written in the form [20]:

$$(K/\rho)M = (K_i/\rho_i)M_i \qquad (4.33)$$

Because $M_i \doteq M$, the above equation implies that K/ρ is approximately a constant for all ice-air mixtures so that K is directly proportional to the snow density [20].

Bohren and Battan [40] emphasize that the validity of the Debye formula is restricted to mixtures with a high degree of homogeneity, and therefore is not applicable to wet snow or "spongy" hail. In their review of several dielectric mixing formulas, Bohren and Battan note that if $(e_{in} - e_m)/3e_m \ll 1$, the formulations of Maxwell-Garnet [41], Bruggemann [42], and Debye [43] give approximately the same result. For ice-water mixtures, however, the Maxwell-Garnet formulation appears to provide the best agreement with measurements if the ice spheres are taken as the inclusions and water as the surrounding medium. If i and w refer to ice and water, respectively, then the dielectric constant, e, for the ice-water mixture is given (according to the Maxwell-Garnet formula) by

$$e = e_w \left[1 + \frac{3f(e_i - e_w)/(e_i + 2e_w)}{1 - f(e_i - e_w)/(e_i + 2e_w)} \right] \tag{4.34}$$

where f is the fractional volume of the ice inclusions.

In contrast to the Debye or the effective medium [42] approaches, the Maxwell-Garnet formulation is asymmetric with respect to the constituents of the mixture. In all cases, the water should be used as the medium with the ice particles as the inclusions. A generalization of the theory to elliptically shaped inclusions has been given by Bohren and Battan [44].

Equation (4.34) is appropriate for spongy hail; for partially melted snowflakes or graupel, however, a large fraction of the volume is air. To compute the dielectric constant for air-ice-water mixtures, a cascade approach has been used [30]: in the first step, the dielectric constant for the ice-water mixture (wet ice) is calculated as shown above; in the second step, presumably, the wet ice is taken as the medium with air inclusions. Using Wiener's formulation, an alternative approach can be used to calculate the dielectric constant for an air-ice-water mixture [31].

4.6 RAYLEIGH APPROXIMATION

In the remote sensing of clouds at microwave frequencies, the diameters of the hydrometeors are much less than the wavelength. This condition is also approximately valid for longer wavelengths ($\lambda > 3$ cm) at lighter rain rates. For spherical particles, the cross sections are given by (see Table 4.2):

$$\sigma_a = \frac{\pi^2}{\lambda} D^3 Im(-K) \tag{4.35}$$

$$\sigma_b = \frac{\pi^5}{\lambda^4} D^6 |K|^2 \tag{4.36}$$

$$\sigma_s = 2/3[(\pi D)^6/\lambda^4]|K|^2 \tag{4.37}$$

$$\sigma_t = \sigma_s + \sigma_a \doteq \sigma_a \tag{4.38}$$

Because the cross sections are related to the drop diameter, D, by a power law, the quantities η, Z, Z_e, k and M in Table 4.2 are moments of the DSD. So too is the rain rate if the velocity distribution is expressed in the form $v = aD^b$ (see Section 4.4). The results are shown in Table 4.5 for three forms of the DSD, where Γ is the complete gamma function.

Table 4.5

Quantity $C_x(x = \eta, Z, k, M, \text{ or } R)$	Multiplicative Constant	Modified Gamma [13] $M_g(m,q,\nu) = N_0\Gamma(\alpha)/|q|\Lambda^\alpha$ $\alpha = (m+1+\nu)/q$	Exponential	Log-Normal [21] $M_l(\sigma,m,\nu) = N_l\exp[m\nu + (\sigma\nu)^2/2]$
η	$\dfrac{\pi^5}{\lambda^4}|K|^2 10^{-6}$	$C_\eta M_g(m,q,\nu = 6)$	$C_\eta N_0 6!/\Lambda^7$	$C_\eta M_l(\sigma,m,\nu = 6)$
Z_c	$\dfrac{|K|^2}{|K_w|^2}\,10^6$	$C_z M_g(m,q,\nu = 6)$	$C_z N_0 6!/\Lambda^7$	$C_z M_l(\sigma,m,\nu = 6)$
k	$\dfrac{0.434\pi^2}{\lambda}\text{Im}(-K)$	$C_k M_g(m,q,\nu = 3)$	$C_k N_0 3!/\Lambda^4$	$C_k M_l(\sigma,m,\nu = 3)$
M	$\pi\rho/6$	$C_M M_g(m,q,\nu = 3)$	$C_M N_0 3!/\Lambda^4$	$C_M M_l(\sigma,m,\nu = 3)$
R	$0.6\pi u$ $(v = aD^b)$	$C_R M_g(m,q,\nu = 3 + b)$	$C_R N_0\Gamma(4 + b)/\Lambda^{4+b}$	$C_R M_l(\sigma,m,\nu = 3 + b)$

A relationship of particular interest is that between k and M. Because both are proportional to the same moment of the DSD, the relationship between the two quantities is independent of it and given by:

$$k = 8.19 \, \text{Im}(-K)M/\lambda\rho \ (\text{dB/km}) \tag{4.39}$$

We note that for $f > 10 \, \text{GHz}$, $\text{Im}(-K)$ for ice is typically two orders of magnitude smaller liquid water so that the attenuation from ice clouds is usually negligible (see Section 4.5).

Other relationships that link the radar quantities (η, Z, Z_e and k) to the meteorological quantities (M and R) depend on the size distribution. The gamma distribution, for example, yields power law relationships involving two free parameters of the *DSD*. For example, if we choose $v(D) = aD^b$ and eliminate Λ in the expressions for Z_e and R and Z_e and M in Table 4.5, then the $Z_e - R$ and $Z_e - M$ relations can be expressed as functions of N_0 and m:

$$Z_e = \frac{|K|^2 \, 10^6 \, N_0 \, \Gamma(7 + m)}{|K_w|^2[1.9 \, aN_0\Gamma(4 + b + m)]^q} \, R^q; \quad q = (7 + m)/(4 + m + b) \tag{4.40}$$

and

$$Z_e = \frac{|K|^2 \, 10^6 \, N_0 \, \Gamma(7 + m)}{|K_w|^2[0.523\rho N_0\Gamma(4 + m)]^p} \, M^p; \quad p = (7 + m)/(4 + m) \tag{4.41}$$

If N_0 is given as a function of m or if m or N_0 is specified, then the various relationships can be expressed in terms of one free parameter of the DSD. For the Marshall-Palmer [21] rain distribution, with $m = 0$ and $N_0 = 8 \times 10^4 \, m^{-3}/\text{cm}^{-1}$, and with an assumed velocity distribution (e.g., $v = 17.67 \, D^{0.67}$), then the power laws are fully specified:

$$Z = 235R^{1.5} \tag{4.42}$$

$$Z - 2 \times 10^4M^{1.75} \tag{4.43}$$

Comprehensive lists of theoretical and experimental power law relationships for rain as functions of frequency, geographical location and storm type have been compiled by Battan [20].

4.7 CLOUDS

Despite the difficulties of radar monitoring of rain from space, current technology offers the opportunity of measuring a large fraction of precipitation that contributes to the global rainfall. The analogous task of detecting and quantifying a significant portion of the global cloud cover is a far more daunting task. In cases of dense, towering cumuliform clouds where the radar reflectivity factor can exceed 0 dBZ, cloud detection capability is highly probable with some of the designs discussed in Section 3.7. On the other hand, high stratus and lenticular clouds with their narrow drop size spectra and low liquid water contents appear to be well outside the range of present radar design capabilities. Ultimately, combinations of millimeter-wave radar and lidar will probably be needed to sense remotely the full range of cloud types.

To try to understand the scope of the radar problem, we have at one extreme the "mother of pearl" cloud, a high, thin lenticular cloud produced by the ascent of moist air over mountains in high latitudes with reflectivity factors, Z, on the order of -60 dB [11,14]. In what might be termed the mid-range of reflectivities are the stratocumulus (-50 dBZ to -22 dBZ), stratus (-50 dBZ to -20 dBZ), nimbostratus (-45 dBZ to -17 dBZ) and the fair-weather cumulus (-37 dBZ to -13 dBZ) [14,45]. At the high end of the reflectivity scale are the cumulus congestus and cumulonimbus, as well as some orographic clouds, characterized by broad droplet spectra and high liquid water contents, that are highly reflective (on the order of 0 dBZ) and highly attenuating. Apart from the low values of reflectivity, the detection problem is further complicated by large spatial variability in cloud liquid water, attenuation both by the cloud water droplets and atmospheric gases, and the relative lack of high power space-qualified transmitters at millimeter wavelengths. Before discussing the question of radar detectability, we will review some of the characteristics of clouds.

One of the reasons for the importance of cloud detection and estimation arises from the fact that a knowledge of the amount and spatial distribution of cloud liquid water provides information required to calculate the growth rates of rain drops [46]. Cloud drop size spectra are of interest not only in the information they convey on the cloud processes but also because the largest drops are most effective in initiating the aggregation process that leads to precipitation. The typical concentration of larger drops needed to initiate precipitation has been estimated to be on the order of $1/m^3$, which constitutes a fraction between about 10^{-5} and 10^{-6} of the the total droplet concentration [28]. The condensation of water vapor and the release of heat to the environmental air is one of the primary forcing mechanisms in large scale circulations in the atmosphere. Clouds not only act to transfer heat from the surface, but also influence the radiational distribution of heating in the ocean and atmosphere. The complex interactions between the circulation patterns of the atmosphere and ocean, and the roles played by clouds and precipitation, are just now becoming understood and quantified. As stated in a recent report: "Clouds tend to heat the atmospheric column through radiative flux convergence . . . The resultant radiational heating gradient between climatologically clear and cloudy regions is at least half the magnitude of the latent heating gradient. While they heat the atmospheric column, clouds also cool the surface below them . . . Clouds also change the proportion of visible to long-wave radiation reaching the surface and, thus, change the radiative heating distribution in the upper layers of the ocean [47]."

Classifications are useful to gain a general idea of the range of heights, temperatures, and vertical air motions that characterize cloud types. In the classification of Mason [46], clouds are grouped into three main categories of stratiform, cumuliform, and special types such as orographic, and irregular, broken clouds. The *stratiform* or layered clouds range from the high-level *cirrus* (above 20,000 ft at temperatures below −25°C) to the *stratocumulus* (below 7000 ft and usually above −5°C) to the rain-bearing *nimbostratus* (often reaching the surface and associated with widespread, uniform vertical air speeds between 5 and 20 cm/s). The *cumuliform* clouds, with their relatively flat bases and clustered towers, range from the fair-weather cumulus with updraft speeds of 1 to 5 m/s to the *cumulonimbus*, with towers exceeding 40,000 ft, tops as cold as −50°C and updrafts that can exceed 30 m/s. These clouds are often associated with cloud electrification, heavy rain rates and hailstorms.

Figure 4.6 schematically shows some of the cloud processes active in the formation of precipitation [48]. While we certainly cannot do justice to the complexity of the growth or interaction processes in clouds, accounts exist in a number of excellent texts on cloud physics [14–15,28,38,46,49]. The primary mechanisms for the growth of hydrometeors from *cloud condensation nuclei* (CCN) (with a typical size of 0.1 micron) to a precipitation-sized particle (on the order of 1 mm) are: (a) diffusion of water vapor to and deposition onto ice crystals; and (b) *accretion*, (a collision between water droplets or ice crystals) followed by *coalescence*—a process which is assisted by turbulence and electrical forces between the hydrometeors. Accretion is by far the most important growth mechanism in warm rain clouds (i.e., clouds whose tops do not exceed the 0°C isotherm) and also plays an essential role in cool clouds (whose tops exceed the 0°C isotherm). In the latter case, however, growth by water vapor deposition onto ice crystals is much more effective than condensation of vapor onto water droplets in warm clouds because the environmental air tends to be supersaturated with respect to ice but only at or near saturation with respect to water [15]. Discussions of the relative importance of growth by accretion and deposition can be found in the literature [15,20,28]. Rogers [15] notes that radar observations often show that the first detectable radar returns of precipitation appear below the 0°C isotherm, suggesting that in these cases the accretion process initiates the precipitation.

Although size distributions of cloud droplets are highly variable, some general properties of these distributions can be enumerated. Because the collisional efficiency increases with the number of large drops, the number concentration of droplets, N_T, is inversely related to average drop size [28]. For example, in the

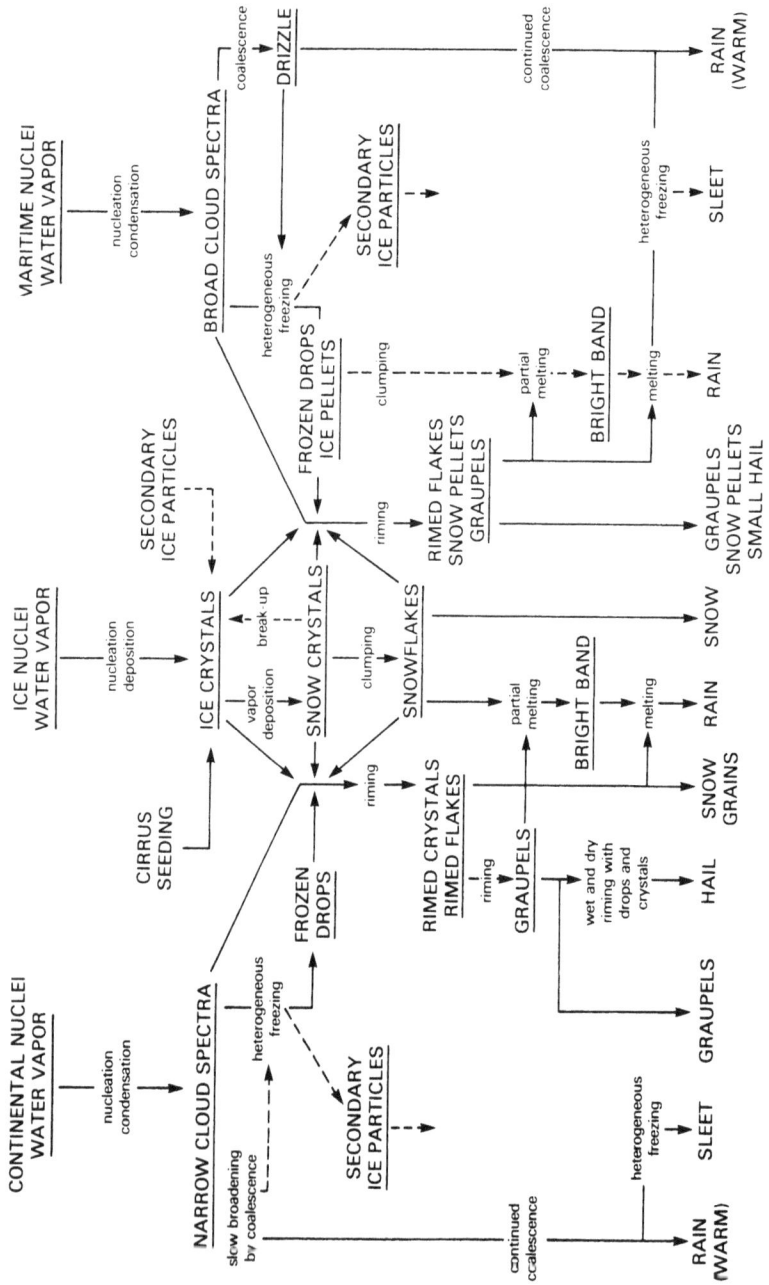

Figure 4.6 Schematic of the growth processes of various types of precipitation (from Braham and Squires [48]).

development from fair-weather cumulus to *cumulus congestus* and cumulonimbus, the increase in the average dropsize is usually accompanied by a decrease in N_T. The number concentration, moreover, increases with vertical air velocity at the cloud base and with decreasing cloud base temperature; N_T also increases with the concentration and size of the CCN [26]. Within the cloud, the liquid water content typically increases above the cloud base and attains a maximum in the upper half of the cloud [37]. Because of the lower concentrations of CCN in maritime regions, N_T tends to be about a factor of six smaller in maritime air than in continental clouds of comparable liquid water content [28,50]. From this it follows that the mean droplet diameter in maritime clouds is a factor of about 1.8 larger than that in continental clouds. This difference has implications in the remote sensing of maritime and continental clouds, because if the LWC is held fixed, an increase by a factor of two in mean droplet diameter will increase the radar reflectivity by about 9 dB. In contrast, the specific attenuation, k, for the two cloud types would be the same for clouds with equal liquid water contents.

Clouds are often characterized by three parameters: the number concentration, $N_T(m^{-3})$, the median volume drop diameter, $D_0(cm)$, and the cloud liquid water content, $M(gm\ m^{-3})$ which can be related to a three-parameter size distribution. In the case of the gamma distribution (see Section 4.3), for example, the parameters (N_0, Λ, m) can be related to $(N_T, D_0,$ and $M)$ by the equations:

$$\frac{\Gamma(m + 1)}{\Gamma(m + 4)} (m + 3.67)^3 = \pi\rho N_T D_0^3/6M \tag{4.44}$$

$$N_0 = N_T(m + 3.67)^{m+1}/\Gamma(m + 1)D_0^{m+1} \tag{4.45}$$

$$\Lambda = (m + 3.67)/D_0 \tag{4.46}$$

From Table 4.5 and (4.44) through (4.46), Z can be expressed as:

$$Z = \frac{6 \times 10^6 \Gamma(m + 7)MD_0^3}{\pi\rho(m + 3.67)^3\Gamma(m + 4)} \ (mm^6\ m^{-3}) \tag{4.47}$$

A numerical solution of (4.44) for m allows Z to be expressed in terms of the given cloud parameters N_T, D_0, and M.

Comparing (4.47) with the empirical $Z - M$ relationship developed by Bartnoff and Atlas [14,51]:

$$Z = 2.58 \times 10^6 D_0^3 M/\rho \ (mm^6\ m^{-3}) \tag{4.48}$$

shows that the two equations agree if $m = 8.7$. Examples of measured size distributions and their characteristics are shown in Figure 4.7 and Table 4.6, respec-

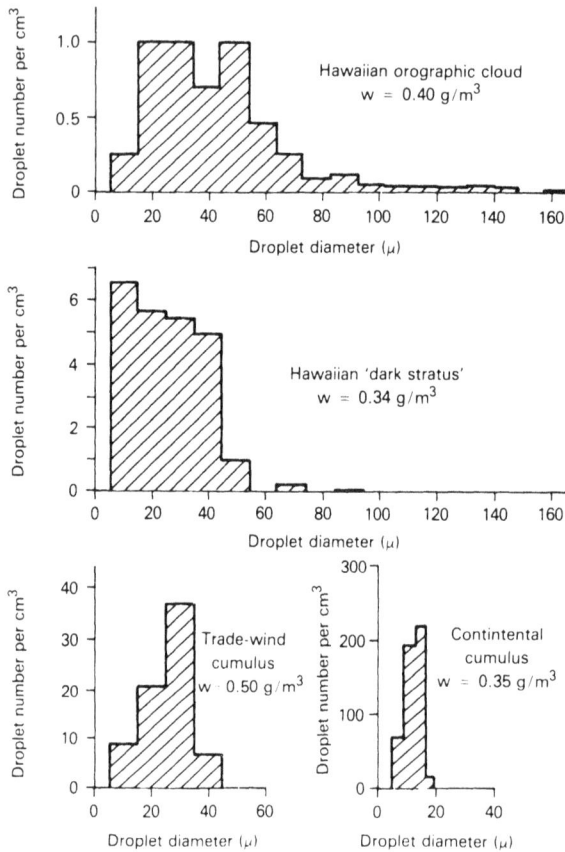

Figure 4.7 Drop spectra of several cloud types of different liquid water contents, w. Cumulus samples are taken 2000 ft above the cloud base, orographic and dark stratus values are average (after Squires [50]).

Table 4.6 Cloud Characteristics and dBZ Values for Several Cloud Types [After 14,50]

Cloud Type	N_T (m^{-3}) ($\times 10^6$)	M (gm m^{-3})	D_0 (cm) ($\times 10^{-4}$)	dBZ (measured)	dBZ ($m = 0$)	dBZ ($m = 4$)	dBZ ($m = 8.7$)	dBZ (4.44)
Continental Cumulus	495	0.35	13.2	−27.4	−24.3	−26.1	−26.7	−26.6 ($m = 6.8$)
Trade-Wind Cumulus	72.5	0.81	26.5	−12.8	−11.5	−13.4	−14.1	—
Hawaiian Dark Stratus	23.3	0.335	34	−14.0	−12.1	−14.0	−14.7	−14.9 ($m = 11$)
Hawaiian Orographic	5.2	0.523	92	1.3	2.8	0.9	0.2	2.0 ($m = 1.1$)

tively. In Table 4.6, the first four columns are taken from the measured data of Squires [48], while the right-hand columns show calculated values of dBZ based on the gamma distribution for $m = 0, 4, 8.7$ (Bartnoff-Atlas) and m as determined by (4.44) and (4.45). For this limited number of examples, the gamma distribution with $m = 4$ is in best agreement with the measurements. In the case of tradewind cumulus, there exists no value of m which satisfies (4.44) and (4.45).

As noted previously, the specific attenuation in the case of Rayleigh scattering is

$$k = 8.19 \, \text{Im}(-K)M/\lambda\rho \, (\text{dB/km})$$

which shows that k is a function of the temperature and phase state of the hydrometeor through the term $\text{Im}(-K)$ (see Table 4.4). While k is directly proportional to the cloud liquid water content, it is independent of the form of the size distribution. Table 4.7 shows values of the specific attenuation (dB/km) normalized to M (gm/m^{-3}) for five frequencies over a range of drop temperatures. Because some uncertainty exists in values of $\text{Im}(-K)$ for ice, the values of specific attenuation should be taken only as approximations. Nevertheless, the results clearly show that attenuation in ice clouds generally can be neglected. In a recent study of doppler weather radar from space, Lhermitte [52] considered three radar designs at frequencies of 15 GHz, 35 GHz and 94 GHz. The radar parameters were selected to yield approximately the same minimum detectable dBZ at the cloud top, ranging from -25 dBZ at 94 GHz to -22 dBZ at 35 GHz. The design parameters are shown in Table 3.4. With these parameters, the 94 GHz radar can attain the same sensitivity as those of the lower frequencies, but with a substantially smaller peak power and antenna size.

Table 4.7 Ratios of Specific Attenuation to Liquid Water Content for Water and Ice

					$k/M[(\text{dB/km})/(\text{gm m}^{-3})]$					
			Water					*Ice*		
					Frequency (GHz)					
T(°C)	*16*	*24*	*35*	*94*	*140*	*16*	*24*	*35*	*94*	*140*
-20	0.45	0.95	1.77	5.41	6.6	.23 E-4	.26 E-4	.28 E-4	.37 E-4	.42 E-4
-10	0.32	0.7	1.36	5.15	6.82	.31 E-4	.34 E-4	.38 E-4	.49 E-4	.54 E-4
0	0.24	0.53	1.05	4.82	7.06	.11 E-3	.12 E-3	.14 E-3	.18 E-3	.21 E-3
10	0.18	0.4	0.82	4.37	7.12					
20	0.14	0.31	0.64	3.85	6.91					
30	0.11	0.25	0.52	3.32	6.45					

Despite these advantages, attenuation is a more severe problem at 94 GHz than at 35 or 15 GHz. For example, the one-way attenuation through a 1 km cloud of liquid water content of 1 gm m^{-3} at 0°C is found in Table 4.8 to be 4.8 dB at 94 GHz, so the minimum detectable reflectivity factor will increase from −25 dBZ at the cloud top to −15.4 dBZ at a 1 km penetration depth. For the same cloud at 35 GHz, the minimum detectable dBZ will increase from −22 dBZ at the storm top to about −20 dBZ at a 1 km depth. Atmospheric absorption will further degrade these results (Figure 4.2), and even more so at higher frequencies. With respect to dynamic range along the entire column for both cloud and rain, the 15 GHz design is the best. The large peak power and antenna size required to accomplish this, however, renders the design impractical. We note that while a minimum dBZ from a spaceborne weather radar of −20 dBZ can be considered quite good by today's standards, this sensitivity would probably enable the detection of less than half the global cloud cover.

Table 4.8

($T = 10°C$)				$Z = aR^b$			$k = aR^b$		
λ(cm)	f(GHz)	m	$\|K\|^2$	a	b	NSE (%)	a	b	NSE (%)
0.32	93.75	3.08 − i1.66	0.759	23	0.71	68	0.84	0.79	35
0.6	50	3.92 − i2.33	0.869	195	1.03	30	0.42	0.96	15
0.86	34.88	4.62 − i2.68	0.899	355	1.26	34	0.22	1.04	10
1.24	24.19	5.48 − i2.9	0.915	443	1.44	40	0.10	1.1	11
1.87	16.04	6.55 − i2.9	0.924	426	1.53	51	0.042	1.13	16
2.4	12.5	7.18 − i2.74	0.927	393	1.53	56	0.023	1.16	20
3.2	9.375	7.81 − i2.43	0.929	360	1.50	58	0.011	1.17	27
6.0	5.	8.68 − i1.58	0.931	341	1.41	41	0.0018	1.05	26

4.8 MIE THEORY

When the particle size is no longer small with respect to the radar wavelength, the cross sections must be solved numerically. Reviews of the theories and their applications to various particle shapes can be found in the literature [8,53–55]. The most important special case is that of scattering by spheres. Treatments of Mie theory [2,20,55–57] and generalizations to concentric spheres [58–60] are numerous. The cross sections for spheres and concentric spheres are in the form of infinite series, the first terms of which yield the Rayleigh solutions. By taking additional terms of the series for the total (or extinction) cross section, σ_t can be represented as a truncated power series in $\pi D/\lambda$. The specific attenuation is then obtained by integrating the series term by term with one of the forms of the DSD discussed earlier. The $k - R$ relationship can be written in the form of a modified power law: $k = \alpha R^{\beta} (1 + e)$, where e represents the effects of non-Rayleigh scattering [23]. A similar technique has been used by Goldstein [61] for attenuation of dry hailstones.

A more common procedure is to employ either analytic forms or measured values of the DSD and compute linear regressions of the form $\log \alpha = a + b \log \beta$ where α is k, Z, Z_e or η, and β is either M or R. For example, tables of $k - R$ relationships are given by Olsen *et al.* [23] for five different drop size distributions of the exponential type for selected frequencies between 1 and 1000 GHz. Figure 4.8 shows averaged measured drop size distributions for various rain rates compared with the Laws and Parsons [6] (L-P) and Marshall-Palmer [21] (M-P) distributions. On the right, the corresponding values of the specific attenuation (dB/km) are shown as a function of frequency for the eight rain rates and three drop size distributions [62]. The most widely used relationship is that between Z and R. Because the relationship is quite sensitive to changes in the drop size distribution, a great deal of variability will occur between individual measurements. One way to visualize the dependence of the Z-R relationship on the DSD parameters is by means of the rain parameter diagram. Such graphical displays have been used not only for rain [18,63–65] but for hail [66] and cloud [14] as well. The example in Figure 4.9 shows the curves corresponding to five Z-R laws, superimposed upon which are curves constant median drop diameter, D_0, and

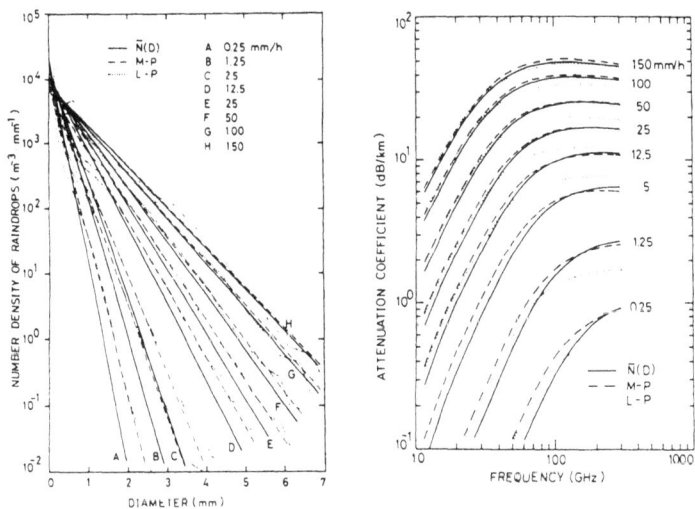

Figure 4.8 (a) Drop size distributions (Laws and Parsons (L-P), Marshall and Palmer (M-P) and inferred) for eight rain rates; (b) corresponding values of the specific attenuation (dB/km) *versus* frequency (from Ihara *et al.* [62]).

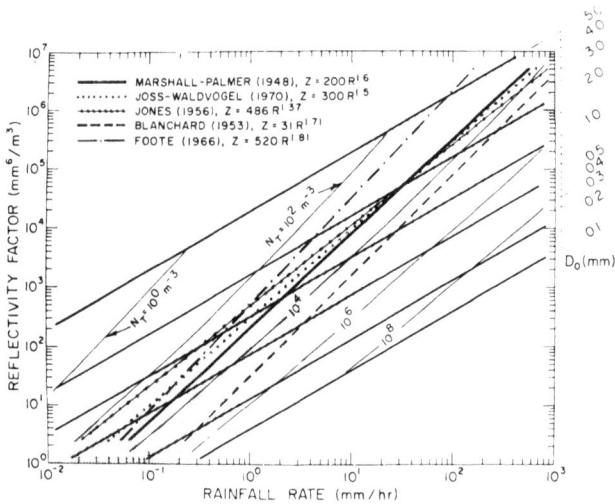

Figure 4.9 Five *Z-R* relationships plotted on the rain parameter diagram with curves of constant drop density (N_T) and curves of constant median volume drop diameter (D_0) (from Ulbrich and Atlas [65]).

number density, N_T. Because an exponential distribution is assumed in this case (two free parameters), a knowledge of Z and either N_T or D_0 uniquely determines R. Of course, it is not only the Z-R or Z_e-R law that is subject to such variations, but virtually all the radar-meteorological relationships. Table 4.8 shows Z-R and k-R relationships for several frequencies based upon measured drop size distributions [35] along with the normalized standard error of the fit. Of particular interest is the minimum in the standard error in the k-R law near 35 GHz. Approximating σ_t by CD^n where C and n are functions of temperature and wavelength, Atlas and Ulbrich [35] showed that $n = 3.67$ for frequencies in the region of 35 GHz. If the velocity-drop diameter relation given by (4.28) is used, it follows that k is proportional to the same moment of the DSD as R. Thus, k and R are linearly related and independent of the DSD. For elevated paths or when updrafts or downdrafts are present, this is no longer strictly valid. Nevertheless, one of the attractive features of attenuation methods at and near 35 GHz is the relative insensitivity of the estimate to fluctuations in the size distribution.

4.9 MELTING LAYER

In stratiform rain, radar observations below the 0°C isotherm often show an enhancement of the return power within a vertical layer on the order of 500 m. This bright-band corresponds to the melting of frozen hydrometeors into rain.

Although the same process occurs in convective storms, due to strong updrafts and mixing within the cloud, supercooled water, frozen aggregates, and partially melted hydrometeors may all be present at levels with temperatures above $-20°C$. In such cases, observations of the melting process may be possible only with the use of polarimetric or dual-wavelength radar. The melting layer or bright-band is pertinent to spaceborne radars because this marker of the 0°C isotherm provides an approximate separation of the precipitation into regions of frozen and melted hydrometeors. Knowledge of the thermodynamic phase is a necessary condition for all quantitative measures of the precipitation. Even in convective systems such as squall lines observed off the Western coast of Africa, stratiform rain accounts for about 25 percent of the total rainfall on average. As the storm ages and the updrafts diminish, the fraction of stratiform rain increases [47,67]. Extrapolating the height of the bright-band into the convective cores provides at least a crude estimate as to the location of mixed-phase hydrometeors. For climatological studies, where heating within the atmosphere is of primary importance, a simple separation of precipitation into convective and stratiform regions may be useful due to the differential heating profiles associated with each [68]. The shape and magnitude of the heating profile and its interaction with the wind field are closely linked to large-scale atmospheric circulations and possibly related to anomalies such as the El-Niño Southern Oscillation phenomenon [47].

Radar observations and modeling of the melting layer have a history almost as long as that of radar meteorology itself [20]. Nevertheless, only in recent years have *in situ* measurement techniques advanced to the stage where detailed studies of the melting process are possible [69–70]. For assemblages of dendrites, the melting first appears as water droplets at branch points near the tips of the crystals. As the particle begins to melt, the melted water tends to flow towards the droplets, leaving large air inclusions in a highly irregular structure. Further growth at the droplet sites and loss of the ice lattice suggest a state just prior to drop break-up. In the later stages of melting, subsequent to drop break-up or collapse, the boundaries of the particles appear smooth and without air inclusions. The shapes, often "lens-like," seem to be nearly independent of mass and of the original crystal type [70]. Observations of the melting process of columnar crystals and ice particles have also been described [69,71].

A large number of models of the melting layer have been proposed and analyzed. By necessity, the models have been highly idealized; nevertheless, they correctly reproduce many of the observable features such as shape and magnitude of the reflectivity profile, wavelength dependence, and doppler signature. It is difficult to assess their quantitative capabilities, however, because of the large number of free parameters in the models. Early studies of the melting layer assumed that the melting particle could be modeled as a sphere composed of a mixture of air, ice and water, where the effective dielectric constant of the mixture was determined by the Debye formula (see Section 4.5). Since the time when the limitations of the Debye formula were pointed out [40], more accurate expressions for the effective dielectric constant have been employed [30–31,72]. Recent models have also included accounts for nonspherical drops, drop break-up, aggregation, and growth by condensation [30–31,72]. An alternative to the inhomogeneous mixture is the concentric sphere model with an inner core of snow surrounded by a water shell [9–10,58,73–74]. Efficient calculation of the Mie coefficients for this geometry can be found in the literature [59–61,75].

For both models, the basic mechanisms that help explain the observations are:

(a) Because the dielectric constant of water is much higher than that of ice, the radar reflectivity increases strongly with the onset of melting. The magnitude of the increase depends on wavelength, the distribution of the melted water on the ice frame, and the snow density.

(b) As the melting proceeds, the air drag decreases, thereby increasing the fall velocities of the particles. Because the smaller particles tend to melt faster, near the top of the melting layer, their velocities increase faster than those of the larger particles. This trend is reversed near the bottom of the layer, where the velocities of the hydrometeors approach those of rain drops, and the velocity is a monotonic function of the drop diameter (see Section 4.4). Because the increase in velocity is accompanied by fewer numbers of particles per unit volume, the reflectivity decreases.

(c) Even if the number density of scatterers is held constant, the backscattering cross section of coated ice or snow-water mixtures generally will be larger than that of a raindrop of equal mass. The amount of increase, however, is dependent upon the initial density of the snow, the distribution of sizes, the radar frequency, and most likely, the details of the melting process.

(d) Increases in the reflectivity result from aggregation of particles (caused by the differential fall velocities and turbulence), and growth by condensation. On the other hand, evaporation, and collisional or spontaneous break-up serve to decrease the reflectivity. For Rayleigh scattering, where the reflectivity is proportional to the sixth power of the diameter, a particle shattering into n equally sized drops leads to a decrease in Z by a factor of $1/n$. The specific attenuation, k, which depends only on the fraction of liquid water per unit volume, should be independent of coalescence or drop break up as long as shape and multiple scattering effects are negligible. In the Mie region, where k is proportional to $D^p(p > 3)$, the break up of the drop will decrease k and aggregation will increase it.

Values of the backscattering cross section for concentric spheres, composed of an inner core of snow of density 0.2 gm/cm^3 and outer shell of water, are shown in Figure 4.10 *versus* the equimass water diameter. The five curves plotted for

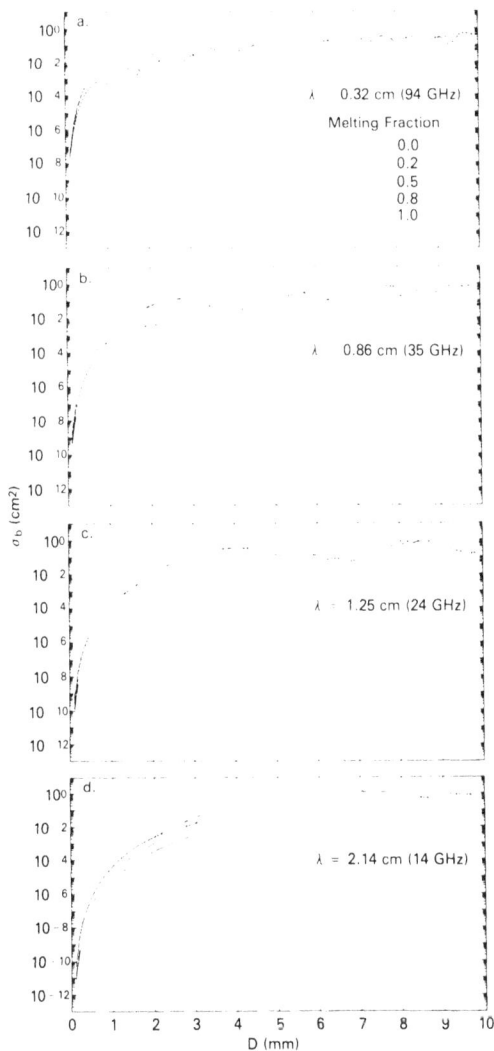

Figure 4.10 Backscattering cross sections *versus* equivolume water diameter for a concentric spherical drop with an inner snow core surrounded by a water shell. The "melting fraction" is the mass of the water to the total mass of the particle: (a) $f = 94$ GHz; (b) 35 GHz; (c) 24 GHz; (d) 14 GHz; (e) 10 GHz (Courtesy of J.A. Jones).

each wavelength correspond to different fractions of the water mass as compared to the total mass. Extremes of 0 and 1 therefore represent uniform spheres of snow and water, respectively. Keeping in mind that the vast majority of drops are smaller than 4 mm, the results imply that the enhancement of the reflectivity in the melting layer relative to that in the snow or rain is greater at the lower frequencies. This difference can be seen in the experimental data shown in Figure 2.12(b) at 10 GHz and 35 GHz. Corresponding curves of the extinction cross section *versus* equivalent water diameter are shown in Figure 4.11. By taking measured drop size

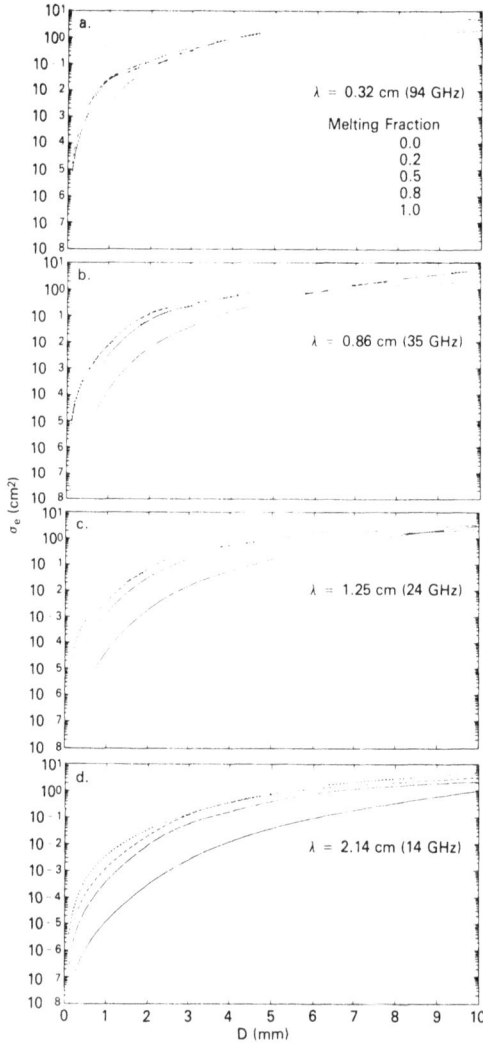

Figure 4.11 Total (extinction) cross section *versus* equivolume water diameter for a concentric spherical drop with an inner snow core surrounded by a water shell. The "melting fraction" is the mass of the water to the total mass of the particle: (a) $f = 94$ GHz; (b) 35 GHz; (c) 24 GHz; (d) 14 GHz; (e) 10 GHz (Courtesy of J.A. Jones).

distributions (DSD) at the surface, converting them to snow spheres of the same mass and letting them melt in accordance with a concentric sphere model without aggregation or break-up [10], qualitative information can be obtained on the dependence of the bright-band on radar wavelength, distribution of drop sizes, snow density and height. The curves in Figure 4.12 show $10 \log Z_e$ versus distance from the 0°C isotherm at several wavelengths for a rain rate of 1 mm/h. The extent of the vertical bars is equal to twice the standard deviation in $10 \log Z_e$. Because growth, evaporation, and particle interactions are not represented in this simple model, the variation in Z_e is caused solely by changes in the drop size distribution.

Using regression equations to relate the calculated values of k, Z_e to R for each drop size distribution for each of several levels, the relationships in Table 4.9 are obtained for initial snow densities of 0.05 and 0.2 gm cm^{-3}. For lower snow densities, the maximum values of reflectivity and attenuation increase and shift to heights closer to the 0°C isotherm. Of course, because of the idealized model of the melting process employed, the results should be used with caution.

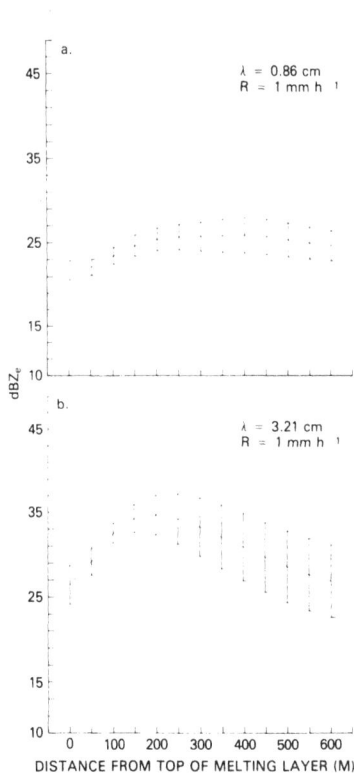

Figure 4.12 Model calculations of Z_e versus range as measured downward from the 0°C level for an initial snow density of 0.2 gm/m^{-3}: (a) f = 35 GHz; (b) 16 GHz; (c) 9.35 GHz; (d) 5.36 GHz (Courtesy of J.A. Jones).

Table 4.9(a) Approximate k-R, Z_e-R, k-Z_e Relationships as a Function of Distance from the Top of the Melting Layer. $\rho_s = 0.2$ gm/cm^3

$\lambda = 0.87$ cm
$T = 0°C$

range (m)	$k = aR^b$		$Z_e = aR^b$		$k = aZ_e^b$	
	a	b	a	b	a	b
0.0	6.223E-02	1.316	1.892E+02	1.092	1.222E-04	1.191
25.0	2.099E-01	0.942	1.921E+02	1.078	2.528E-03	0.845
50.0	4.719E-01	0.827	2.082E+02	1.030	7.348E-03	0.783
75.0	7.359E-01	0.822	2.412E+02	0.969	7.475E-03	0.838
100.0	9.310E-01	0.862	2.837E+02	0.927	5.106E-03	0.923
150.0	1.091E+00	0.972	3.761E+02	0.931	2.429E-03	1.032
200.0	1.027E+00	1.079	4.444E+02	1.008	1.633E-03	1.059
250.0	8.602E-01	1.172	4.697E+02	1.109	1.386E-03	1.047
300.0	6.915E-01	1.237	4.788E+02	1.201	1.284E-03	1.021
400.0	4.548E-01	1.267	4.920E+02	1.310	1.293E-03	0.949
500.0	3.147E-01	1.220	4.348E+02	1.332	1.403E-03	0.895
600.0	2.564E-01	1.157	3.735E+02	1.313	1.596E-03	0.861
700.0	2.367E-01	1.109	3.561E+02	1.294	1.757E-03	0.838
800.0	2.268E-01	1.076	3.464E+02	1.275	1.866E-03	0.824
900.0	2.227E-01	1.058	3.440E+02	1.269	1.949E-03	0.815
1000.0	2.206E-01	1.049	3.430E+02	1.266	1.994E-03	0.810
1100.0	2.198E-01	1.045	3.421E+02	1.264	2.009E-03	0.808
1200.0	2.195E-01	1.044	3.416E+02	1.263	2.012E-03	0.808
1300.0	2.195E-01	1.044	3.415E+02	1.262	2.013E-03	0.808
1400.0	2.195E-01	1.044	3.415E+02	1.262	2.013E-03	0.808

$\lambda = 3$ cm
$T = 0°C$

range (m)	$k = aR^b$		$Z_e = aR^b$		$k = aZ_e^b$	
	a	b	a	b	a	b
0.0	1.969E-03	1.165	5.527E+02	1.460	1.441E-05	0.782
25.0	6.617E-02	0.711	6.697E+02	1.406	3.082E-03	0.468
50.0	1.617E-01	0.744	1.011E+03	1.310	4.103E-03	0.536
75.0	2.343E-01	0.819	1.546E+03	1.249	2.322E-03	0.632
100.0	2.653E-01	0.904	2.186E+03	1.237	1.109E-03	0.714
150.0	2.405E-01	1.070	3.282E+03	1.300	3.416E-04	0.811
200.0	1.753E-01	1.218	3.621E+03	1.415	1.692E-04	0.849
250.0	1.183E-01	1.339	3.266E+03	1.539	1.186E-04	0.856
300.0	8.044E-02	1.421	2.626E+03	1.650	1.059E-04	0.845
400.0	4.338E-02	1.467	1.528E+03	1.779	1.194E-04	0.807
500.0	2.756E-02	1.440	9.018E+02	1.803	1.366E-04	0.783
600.0	2.081E-02	1.383	6.244E+02	1.775	1.559E-04	0.764
700.0	1.786E-02	1.316	5.196E+02	1.729	1.757E-04	0.743
800.0	1.594E-02	1.243	4.582E+02	1.677	1.993E-04	0.720
900.0	1.487E-02	1.182	4.193E+02	1.616	2.157E-04	0.706
1000.0	1.431E-02	1.143	3.926E+02	1.556	2.146E-04	0.709
1100.0	1.409E-02	1.127	3.782E+02	1.516	2.032E-04	0.719
1200.0	1.403E-02	1.122	3.724E+02	1.498	1.950E-04	0.727
1300.0	1.402E-02	1.122	3.711E+02	1.495	1.928E-04	0.729
1400.0	1.402E-02	1.122	3.711E+02	1.495	1.928E-04	0.729

Table 4.9(b) Approximate k-R, Z_e-R, k-Z_e Relationships as a Function of Distance from the Top of the Melting Layer. $\rho_s = 0.05$ gm/cm³

$\lambda = 0.87$ cm, $T = 0°C$

range (m)	$k = aR^b$ a	$k = aR^b$ b	$Z = aR^b$ a	$Z = aR^b$ b	$k = aZ_e^b$ a	$k = aZ_e^b$ b
0.0	6.314E-02	1.226	8.319E+01	0.861	1.726E-04	1.347
25.0	5.236E-01	0.769	8.853E+01	0.823	8.877E-03	0.913
50.0	1.382E+00	0.741	1.286E+02	0.735	1.119E-02	0.993
75.0	2.034E+00	0.808	2.152E+02	0.738	6.635E-03	1.069
100.0	2.342E+00	0.893	3.356E+02	0.808	4.344E-03	1.084
150.0	2.121E+00	1.065	5.530E+02	0.973	2.281E-03	1.088
200.0	1.485E+00	1.217	4.698E+02	1.117	1.564E-03	1.086
250.0	9.505E-01	1.327	4.486E+02	1.238	1.312E-03	1.071
300.0	6.417E-01	1.354	3.967E+02	1.307	1.213E-03	1.028
400.0	3.337E-01	1.280	3.589E+02	1.332	1.183E-03	0.946
500.0	2.544E-01	1.169	3.494E+02	1.306	1.478E-03	0.878
600.0	2.308E-01	1.094	3.453E+02	1.284	1.796E-03	0.833
700.0	2.224E-01	1.057	3.427E+02	1.265	1.972E-03	0.812
800.0	2.201E-01	1.046	3.415E+02	1.262	2.006E-03	0.808
900.0	2.195E-01	1.044	3.415E+02	1.262	2.013E-03	0.808
1000.0	2.195E-01	1.044	3.415E+02	1.262	2.013E-03	0.808
1200.0	2.195E-01	1.044	3.415E+02	1.262	2.013E-03	0.808
1300.0	2.195E-01	1.044	3.415E+02	1.262	2.013E-03	0.808
1400.0	2.195E-01	1.044	3.415E+02	1.262	2.013E-03	0.808

$\lambda = 3$ cm, $T = 0°C$

range (m)	$k = aR^b$ a	$k = aR^b$ b	$Z = aR^b$ a	$Z = aR^b$ b	$k = aZ_e^b$ a	$k = aZ_e^b$ b
0.0	3.159E-03	1.137	7.793E+02	1.386	1.509E-05	0.805
25.0	2.376E-01	0.683	1.236E+03	1.258	6.229E-03	0.515
50.0	6.210E-01	0.744	2.856E+03	1.102	3.165E-03	0.665
75.0	8.368E-01	0.854	5.588E+03	1.074	8.993E-04	0.792
100.0	8.649E-01	0.968	8.626E+03	1.117	3.497E-04	0.863
150.0	6.121E-01	1.198	1.167E+04	1.293	1.165E-04	0.916
200.0	3.346E-01	1.414	9.752E+03	1.511	7.487E-05	0.917
250.0	1.740E-01	1.572	6.229E+03	1.721	7.270E-05	0.894
300.0	9.736E-02	1.631	3.645E+03	1.853	8.715E-05	0.860
400.0	3.774E-02	1.609	1.281E+03	1.935	1.095E-04	0.819
500.0	2.311E-02	1.482	7.070E+02	1.879	1.444E-04	0.776
600.0	1.756E-02	1.332	5.262E+02	1.784	1.887E-04	0.728
700.0	1.505E-02	1.198	4.313E+02	1.655	2.267E-04	0.697
800.0	1.422E-02	1.137	3.882E+02	1.547	2.163E-04	0.708
900.0	1.402E-02	1.122	3.711E+02	1.495	1.928E-04	0.729
1000.0	1.402E-02	1.122	3.711E+02	1.495	1.928E-04	0.729
1200.0	1.402E-02	1.122	3.711E+02	1.495	1.928E-04	0.729
1300.0	1.402E-02	1.122	3.711E+02	1.495	1.928E-04	0.729
1400.0	1.402E-02	1.122	3.711E+02	1.495	1.928E-04	0.729

REFERENCES

[1]   Waters, J.W., 1976: Absorption and emission of microwave radiation by atmospheric gases, in Methods of Experimental Physics, M.L. Meeks, ed. 12, Part B, *Radio Astronomy,* Academic Press, Section 2.3.

[2]   Ulaby, F.T., R.K. Moore, and A.K. Fung, 1981: *Microwave Remote Sensing: Active and Passive.* Vol. I. Artech House, Norwood, MA, 456 pp.

[3]   Rosenkranz, P.W., 1975: Shape of the 5 mm oxygen band in the atmosphere. *IEEE Trans. Ant. and Propag.,* **AP-23,** 498–506.

[4]   Crane, R.K., and D.W. Blood, 1979: Handbook for the estimation of microwave propagation effects-link calculations for earth-space paths. Rep. P-7376-TR1, 80 pp., Environmental Research and Technology, Inc, Concord, MA.

[5]   Bean, B.R., and E.J. Dutton, 1966: *Radar Meteorology,* NBS Monograph 92, US Department of Commerce, National Bureau of Standards, Boulder, CO, US Government Printing Office, March.

[6]   Laws, J.O., and D.A. Parsons, 1943: The relation of raindrop-size to intensity. *Trans. Amer. Geophys. Union,* **24,** 452–460.

[7]   Smith, P.L., 1984: Equivalent radar reflectivity factors for snow and ice particles. *J. Clim. and Appl. Meteor.,* **23,** 1258–1260.

[8]   Mon, J.P., 1982: Backward and forward scattering of microwaves by ice particles: A review. *Radio Sci.,* **17,** 953–971.

[9]   Ekpenyong, B.E., and R.C. Srivastava, 1970: Radar characteristics of the melting layer—a theoretical study. Tech. Note No. 16, Lab. Atmos. Probing, University of Chicago, 34 pp.

[10] Yokoyama, T., and H. Tanaka, 1984: Microphysical processes of melting snowflakes detected by two-wavelength radar. Part I. Principle of measurement based on model calculations. *J. Meteor. Soc. of Japan,* **62,** 650–666.

[11] Deirmendjian, D., 1969: *Electromagnetic Scattering on Spherical Polydispersions.* Amer. Elsevier Publishing Co., Inc., New York, 290 pp.

[12] Sekon, R.S., and R.C. Srivastava, 1971: Doppler radar observations of drop-size distributions in a thunderstorm. *J. Atmos. Sci.,* **28,** 983–994.

[13] Ulbrich, C.W., 1983: Natural variations in the analytical form of the raindrop size distribution. *J. Clim. and Appl. Meteor.,* **22,** 1764–1775.

[14] Gossard, E.E., and R.G. Strauch, 1983: *Radar Observation of Clear Air and Clouds.* Elsevier, Amsterdam, 280 pp.

[15] Rogers, R.R., 1979: *A Short Course in Cloud Physics.* Pergamon Press, Oxford, 235 pp.

[16] Sulakvelidze, G.K., and Y.A. Dadali, 1971: Multiwavelength radar measurements of precipitation intensity. Radar Meteorology, *Proc. Third All-Union Conf.* (translated from Russian), Israel Program for Scientific Translations, 32–45 (available from NTIS).

[17] Wong, R.K.W., N. Chidambaram, L. Cheng, and M. English, 1988: The sampling variations of hailstone size distributions. *J. Appl. Meteor.,* **27,** 254–260.

[18] Atlas, D., C.W. Ulbrich, and R. Meneghini, 1984: The multiparameter remote measurement of rainfall. *Radio Sci.,* **19,** 3–22.

[19] Ulbrich, C.W., 1986: A review of the differential reflectivity technique of measuring rainfall. *IEEE Trans Geosci. and Remote Sensing,* **GE-24,** 955–965.

[20] Battan, L.J., 1973: *Radar Observations of the Atmosphere,* University of Chicago Press, Chicago, 324 pp.

[21] Marshall, J.S., and W.M.K. Palmer, 1948: The distribution of raindrops with size. *J. Meteor.,* **5,** 165–166.

[22] Joss, J., C. Thams, and A. Waldvogel, 1968: The variation of raindrop size distributions at Locarno, *Proc. Int. Conf. Cloud Physics,* pp. 369–373.

[23]   Olsen, R.L., D.V. Rogers, and D.B. Hodge, 1978: The aR relation in the calculation of rain attenuation. *IEEE Trans. Ant. and Propag.*, **26**, 318–329.

[24]   Feingold, G., and Z. Levin, 1987: Application of the lognormal distribution to differential reflectivity measurement (ZDR). *J. Atmos. and Oceanic Technol.*, **4**, 377–382.

[25]   Fang, D.J., and C.H. Chen, 1982: Propagation of centimeter/millimeter waves along a slant path through precipitation. *Radio Sci.*, **17**, 989–1005.

[26]   Ajayi, G.O., and R.L. Olsen, 1985: Modeling of a tropical raindrop size distribution for microwave and millimeter wave applications, *Radio Sci.* **20**, 193–202.

[27]   Beard, K.V., 1976: Terminal velocity and shape of cloud and precipitation drops aloft. *J. Atmos. Sci.*, **33**, 851–864.

[28]   Houghton, H.G., 1985: *Physical Meteorology*. MIT Press, Cambridge, MA, 442 pp.

[29]   Gunn, K.L.S., and J.S. Marshall, 1958: The distribution with size of aggregate snowflakes. *J. Meteor.*, **15**, 452–461.

[30]   Klaassen, W., 1988: Radar observations and simulation of the melting layer of precipitation. *J. Atmos. Sci.*, **45**, 3741–3753.

[31]   Awaka, J., Y. Furuhama, M. Hoshiyama, and A. Nishitsuji, 1985: Model calculations of scattering properties of spherical bright-band particles made of composite dielectrics. *J. Radio Res. Lab.*, **32**, 73–87.

[32]   Gunn, K.L.S., and G.D. Kinzer, 1949: The terminal velocity of fall for water droplets in stagnant air. *J. Meteor.*, **6**, 243–248.

[33]   Foote, G.B., and P.S. du Toit, 1969: Terminal velocity of raindrops aloft. *J. Appl. Meteor.*, **8**, 245–253.

[34]   Spilhaus, A.F., 1948: Drop size intensity and radar echo of rain. *J. Meteor.*, **5**, 161–164.

[35]   Atlas, D. and C.W. Ulbrich, 1977: Path- and area-integrated rainfall measurement by microwave attenuation in the 1–3 cm band. *J. Appl. Meteor.*, **16**, 1322–1331.

[36]   Lhermitte, R., 1988: Cloud and precipitation sensing at 94 GHz. *IEEE Trans. Geosci. and Remote Sens.*, **26**, 207–216.

[37]   Douglas, R.H., 1963: Recent hail research: A review. *Meteor. Monographs*, **5**, 157–167.

[38]   Pruppacher, H.R., and K.V. Beard, 1970: A wind tunnel investigation of the internal circulation and shape of water drops falling at terminal velocity in air. *Quart. J. Roy. Meteor. Soc.*, **96**, 247–256.

[39]   Ray, P.S., 1972: Broadband complex refractive indices of ice and water. *Appl. Optics*, **11**, 1836–1844.

[40]   Bohren, C.F., and L.J. Battan, 1980: Radar backscattering by inhomogenous particles. *J. Atmos. Sci.*, **37**, 1821–1827.

[41]   Maxwell Garnett, J.C., 1904: Colors in metal glasses and in metallic films. *Phil. Trans. Roy. Soc. London*, **A203**, 385–420.

[42]   Bruggeman, D.A.G., 1935: Berechnung vershiedener physikalischer Konstanten von heterogenen Substanzen. I, Dielektrizitatskonstanten und Leitfahigkeiten der Mischkorper aus isotropen Substanzen. *Ann. Phys. Leipzig*, **24**, 636–679.

[43]   Debye, P., 1929: *Polar Molecules*. The Chemical Catalog Company, New York, 172 pp.

[44]   Bohren, C.F., and L.J. Battan, 1982: Radar backscattering of microwaves by spongy ice spheres. *J. Atmos. Sci.*, **39**, 2623–2628.

[45]   Boucher, R.J., 1952: Empirical relationship between radar reflectivity drop size distribution and liquid water content in clouds. Mount Washington Observatory, Contract AF 19(122)-399, 14 pp.

[46]   Mason, B.J., 1975: *Clouds, Rain and Rainmaking*. 2nd Ed., Cambridge University Press, London, 189 pp.

[47]   Simpson, J., ed., 1988: TRMM—A satellite mission to measure tropical rainfall. Report of the Science Steering Group. NASA/GSFC, August, 94 pp.

[48]   Braham, R.R., and P. Squires, 1974: Cloud Physics—1974. *Bull. Amer. Meteor. Soc.*, **55**, 543–556.

[49]   Fletcher, N.H., 1962: *The Physics of Rainclouds.* Cambridge University Press, London, 386 pp.

[50]   Squires, P., 1958: The microstructure and colloidal stability of warm clouds. Part I: The relationship between structure and stability. *Tellus*, **10**, 256–261.

[51]   Bartnoff, S., and D. Atlas, 1951: Microwave determination of particle-size distribution. *J. Meteor.*, **8**, 130–131.

[52]   Lhermitte, R., 1989: Satellite-borne millimeter wave Doppler radar. URSI Commission F, Open Symposium, September 11–15, La Londe-Les-Maures, France.

[53]   Oguchi, T., 1981: Scattering from hydrometeors: A survey. *Radio Sci.*, **16**, 691–730.

[54]   Holt, A.R., 1982: The scattering of electromagnetic waves by single hydrometeors, *Radio Sci.*, **17**, 929–945.

[55]   Van de Hulst, H.C., 1957: *Light Scattering by Small Particles.* John Wiley and Sons, New York, 470 pp.

[56]   Born, M., and E. Wolf, 1980: Principles of Optics, 6th Ed. Pergamon Press, Oxford, 808 pp.

[57]   Ishimaru, A., 1978: *Wave Propagation and Scattering in Random Media.* Academic Press, New York, Vol. 1, 250 pp.

[58]   Aden, A.L., and M. Kerker, 1951: Scattering of electromagnetic waves by two concentric spheres. *J. Appl. Phys.*, **22**, 1242–1246.

[59]   Kerker, M., 1969: *The Scattering of Light and Other Electromagnetic Radiations,* Academic Press, New York, 666 pp.

[60]   Toon, O.B., and T.P. Ackerman, 1981: Algorithms for the calculation of scattering by stratified spheres. *Appl. Optics,* **20**, 3657–3660.

[61]   Goldstein, H., 1951: Attenuation by Condensed Water, in Propagation of Short Radio Waves, E.D. Kerr, ed., pp 671–692, McGraw-Hill, New York.

[62]   Ihara, T., Y. Furuhama, and T. Manabe, 1984: Inference of raindrop size distribution from rain attenuation statistics at 12, 35, and 82 GHz. *Trans. IECE of Japan,* **E67**, 211–217.

[63]   Atlas, D., and A.C. Chmela, 1957: Physical-synoptic variations of drop-size parameters. *Proc. Sixth Weather Radar Conf.,* Amer. Meteor. Soc., Boston, pp. 21–30.

[64]   Atlas, D., and C.W. Ulbrich, 1974: The physical basis for attenuation-rainfall relationships and the measurement of rainfall parameters by combined attenuation and radar methods. *J. Rech. Atmos.,* **8**, 275–298.

[65]   Ulbrich, C.W., and D. Atlas, 1978: The rain parameter diagram: Methods and applications. *J. Geophys. Res.,* **83**, 1319–1325.

[66]   Ulbrich, C.W., and D. Atlas, 1982: Hail parameter relations: A comprehensive survey. *J. Appl. Meteor.,* **21**, 22–43.

[67]   Tao, W.K., and J. Simpson, 1984: Cloud interactions and merging: numerical simulation. *J. Atmos. Sci.,* **41**, 2901–2917.

[68]   Houze, R.A., Jr., 1982: Cloud clusters and large-scale vertical motions in the tropics. *J. Meteor. Soc. Japan,* **60**, 396–410.

[69]   Knight, C.A., 1979: Observations of the morphology of melting snow. *J. Atmos. Sci.,* **36**, 1123–1130.

[70]   Fujiyoshi, Y., 1986: Melting snowflakes. *J. Atmos. Sci.,* **43**, 307–311.

[71]   Rasmussen, R.M., V. Levizzani and H.R. Pruppacher, 1984: A wind tunnel and theoretical study on the melting behavior of atmospheric ice particles. Part III: Experiments and theory for spherical ice particles of radius > 500 m. *J. Atmos. Sci.,* **41**, 381–388.

[72]   Bringi, V.N., R.M. Rasmussen, and J. Vivekanandan, 1986: Multiparameter radar measurements in Colorado convective storms. Part I: Graupel melting studies. *J. Atmos. Sci.,* **43**, 2545–2563.

[73] Dissanayake, A.W., and N.J. McEwan, 1978: Radar and attenuation properties of rain and bright band. *IEEE Conf. Publ.* **169-2,** 125–129.

[74] Yokoyama, T., H. Tanaka, K. Nakamura and J. Awaka, 1984: Microphysical processes of melting snowflakes detected by a two wavelength radar. Part II. Application of a two wavelength radar technique. *J. Meteor. Soc. Japan,* **62,** 668–677.

[75] Rumpl, W.M., 1982: Program description and user's guide of subroutine Mie. CSC/TM-82/6193, Computer Sciences Corporation, Contract NAS 5-24350.

Chapter 5
Estimation Methods

Although many ground-based methods are applicable to spaceborne weather radar, there are several differences that influence performance. The most obvious of these is geometry: even intense convective storms rarely exceed altitudes of 18 km. For most designs, this implies that the one-way path through the hydrometeors will be less than 30 km, as compared with a 200 km radar range typical in ground-based systems. Another distinction between ground and space-based weather radars is the presence in the latter of a strong surface return. Although it can mask the rain return near the surface, the surface return can also be used as a reference by which the total path attenuation can be deduced. The surface also gives rise to a number of multiple scattering terms, at least one of which, the mirror-image return, can be substantial. While this component may require a lowering of the PRF to prevent range ambiguity, the mirror-image return may have applications in the estimation of doppler velocities and attenuation. A third difference arises from the size and cost constraints of spaceborne antennas, which favor the use of higher frequencies, typically 10 GHz and above. Although the path lengths through the rain will be shorter, we can generally say that attenuation effects, and therefore attenuation methods, will be of greater importance from space.

In addition to these differences, there are two important meteorological methods that require modifications for their use from space. Because the maximum off-nadir angle probably will be less than 45°, the sensitivity of most polarimetric measurements to highly aligned hydrometeors (such as raindrops) will be reduced. Even if higher incidence angles are used, unless the cross-track beam is very narrow, a variety of hydrometeors typically will be present in a single pulse volume, increasing the difficulty in the estimation. Difficulties are also present in using doppler measurements from space. Although estimation of the mean doppler velocity of the hydrometeors appears possible, even this measurement presents formidable technological challenges, especially in the case of a scanning radar. As discussed in Chapter 2, some of the radar requirements are a high resolution beamwidth in the along-track direction, a high pulse repetition rate, and accurate knowledge and control of the pointing angle.

5.1 BACKSCATTERING METHODS

When the pulse volume is uniformly filled with hydrometeors and when either the rain rate is light or the frequency is low so that attenuation can be neglected, the meteorological radar equation simplifies to

$$P(r) = C|K_w|^2 Z_e/r^2 \tag{5.1}$$

where r is the radar range, C is the radar constant and Z_e is the equivalent reflectivity factor. From this equation and a knowledge of C and r, Z_e can be obtained from an estimate of the mean return power, P, at each range gate.

In listing some of the error sources in the determination of Z_e, we recall that the variance of Z_e is inversely proportional to the number of independent samples that comprise the estimate of P. This is valid for linear, square law and logarithmic receivers if the number of independent samples is greater than about 30. The separation of the radar equation in (5.1) into a product of meteorological and radar terms, however, requires an assumption as to the distribution of hydrometeors within the pulse volume. When the actual distribution differs from that which is assumed, C will be incorrect and Z_e will be biased. A further source of error arises from the effects of attenuation where the use of (5.1) will underestimate Z_e. However, if the receiver noise power or the surface return is a substantial fraction of the return from the hydrometeors, then the estimate of Z_e deduced from the total power will be positively biased.

From Z_e the rain rate, R, or the liquid water content, M, is inferred from power law relationships, as discussed in Chapter 4. The functional relationship between Z_e and R depends, in rough order of importance, on the phase state of the hydrometeors, the drop size and velocity distributions of the particles, shape and orientation effects, and drop temperature. In stratiform rain, the melting layer is normally confined to a vertical region on the order of 500 m, and can be readily identified at lower frequencies as an enhancement of the radar return. Because of Mie scattering effects, the detectability of this bright-band decreases with increasing frequency. Despite the fact that partially melted hydrometeors will occasionally be present well below the 0°C isotherm, the bright-band provides the best means for a single-wavelength radar to distinguish between regions of frozen hydrometeors and rain. In convective rain, the situation is much more difficult, and regions may exist where ice, liquid and partially melted drops are present in the same pulse volume. Without a second independent measurement, recourse to a storm model or auxiliary data sets must be made to estimate the relative concentrations of rain, snow and partially melted hydrometeors. A crude type of categorization can be accomplished by detecting the melting layer in the stratiform portions of the storm and then extrapolating into the convective regions.

Even when the hydrometeors are spherical raindrops, the effects of fluctuations in the drop size distribution can be significant. As noted in Chapter 4, the relationship between Z and liquid water content M can be found by specifying the drop size distribution and the drop temperature. In the case of rain rate, a velocity distribution of drops must be given as well. At the minimum, the DSD requires two independent parameters: the exponential distribution, for example, can be determined by the number density and the median volume drop diameter. Conversely, the standard backscattering method only provides a single measurement. Some reduction in the scatter between Z and R can be attained, however, if the relationships are conditioned on geographical location, rain type, and synoptic conditions. Even with such precautions, instantaneous errors in the Z-R relations can exceed 100 percent [1–2]. Of course, the errors depend on the spatial and temporal scales of the measurement. For example, the Marshall and Palmer distribution provides a good representation of the DSD in stratiform precipitation if large spatial or temporal averages are used [3].

We mention that most Z-R ($Z = aR^b$) relationships are obtained by determining the coefficients a' and b in the equation:

$$\log Z = a' + b \log R; \quad a' = \log a$$

by means of a standard linear regression. Because the bulk of the measurements occur at light rain rates, this procedure tends to highly weight these data. To obtain Z-R or R-Z relationships that are more accurate at moderate and high rain rates (i.e., those rain rates that contribute most to the total rainfall), we can proceed either by using piece-wise linear regressions or by introducing a weighting function. To outline this latter possibility, we note the coefficients a and b in the regression equation are obtained by minimizing the quantity E:

$$E = \sum_{i=1}^{N} w(y_i)[y_i - (a + bx_i)]^2$$

where $w(y_i)$ is the weighting function, and a and b are found by setting the partial derivatives of E with respect to a and b equal to zero. In most cases, the weighting function is taken to be unity. To increase the influence of the higher rain rates, the weighting function can be taken to be y_i^m with $m = 1$ or 2. The parameters a and b are again determined by setting the partial derivatives of E, with respect to a and b, equal to zero.

In principle at least, the weighted regression is similar to the distribution matching procedure that has been investigated recently [4–6]. Letting $p(R)$ and $p(Z)$ represent the probability density functions for the rain rate and reflectivity factor, respectively, then the nine pairs of R, Z values can be determined from the equations [5,6]:

$$\frac{\int_0^{R_i} R^j p(R)\, dR}{\int_0^{\infty} R^j p(R)\, dR} = 0.1i; \quad \frac{\int_0^{Z_i} Z^j p(Z)\, dZ}{\int_0^{\infty} Z^j p(Z)\, dZ} = 0.1i \tag{5.2}$$

where $i = 1, \ldots, 9$ and $j = 0, 1,$ or 2.

Using $j = 1$ and plotting pairs of (Z_i, R_i) values on a log Z *versus* log R graph and performing a linear regression yields the two parameters in the Z-R relationship. Differences in the Z-R relationships derived from this and the usual method can be substantial [6], which suggests that a single Z-R power law may not always be representative throughout the full span of rain rates. One of the attractive features of the matching procedure is that the distributions of R (generally derived from a rain gauge network) and Z (derived from the radar) need not be made simultaneously as long as the data bases comprising R and Z are representative of a particular climatology [6]. A problem with all relationships between radar derived quantities and the rainfall rate is that the former are independent and the latter dependent on the velocity distribution of the hydrometeors. The deviations from the assumed distribution that occur, for example, in the presence of updrafts and downdrafts, introduce errors in the rain rate estimate. This is the case for backscattering, attenuation, and polarimetry methods. Despite the many problems with the backscattering method, it continues to be the most commonly used of all meteorological methods. Its advantages from both ground and space-based platforms are simplicity, range-profiling capabilities, and large dynamic range.

5.2 SINGLE-WAVELENGTH ATTENUATION METHODS

If the attenuation is not negligible, the radar equation becomes

$$P(r) = C|K_w|^2 Z_e(r) \exp\left[-0.2 \ln 10 \int_0^r k(s)\, ds\right]/r^2 \tag{5.3}$$

For a single frequency and polarization, there exist two unknown quantities, $Z_e(r)$ and $k(r)$ for each measurement of power. In the simplest case, where the characteristics of the rain (or cloud) can be assumed uniform with range, a plot of $5 \log 10[P(r)]$ *versus* r will have a slope equal in magnitude to the specific attenuation k (dB/km) [7]. From k, the rainfall rate or liquid water content is obtained by means of power law relations between k and M or k and R. A technique of this type has been considered for spaceborne radar applications using the concept of "early range" gates on altimeters such as SEASAT [8]. Plots of the signal power (in dBm) are shown in Figure 5.1 for various depths of penetration into the storm as a function of rainfall rate.

In general, the properties of the rain and cloud will change as a function of distance. When the fluctuations in the reflectivity factor within a distance d become comparable to the attenuation over d, then the uniformity assumption is not valid. In these cases, an alternative is to reduce the number of unknown parameters by relating k to Z_e by a power law of the form $k = \alpha Z_e^\beta$. Substituting this into (5.3) allows $P(r)$ to be expressed as a function of $Z_e(s)$ ($s < r$). To invert this equation so that $Z_e(r)$ can be written as a function of $P(s)$ ($s < r$), the equation can be transformed into a differential equation of the Bernoulli type. Solving this and using $R = aZ_e^b$ yields an estimate for the rain rate at range r:

$$R(r) = aZ_m^b(r)[p(r)]^{-b/\beta} \tag{5.4}$$

with

$$p(r) = 1 - 0.2 \ln 10\beta \int_0^r \alpha Z_m^\beta \, ds \tag{5.5}$$

where the measured, or apparent, reflectivity factor is defined by:

$$Z_m(r) = r^2 P(r)/C|K_w|^2 \tag{5.6}$$

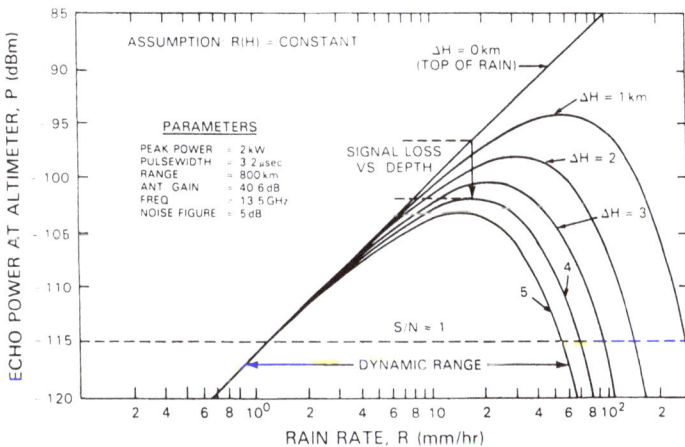

Figure 5.1 Radar return power *versus* range for various depths within the storm (from Goldhirsh and Walsh [8]).

We should mention that the definition of p given by (5.5) is an estimate of the quantity $\exp(-0.2 \ln 10 \beta \int_0^r k(s) \, ds)$ in terms of the measured values $Z_e(s)$, $s < r$. This follows directly from a comparison of (5.4) with the equations

$$R(r) = aZ_e^b(r) = aZ_m^b(r) \exp\left[0.2 \ln 10b \int_0^r k(s) \, ds\right]$$

The parameters a, b, α and β are obtained from an assumed DSD, temperature, and frequency using Mie theory. Because of the restrictions on the derivation, although a, b, and α are allowed to vary with range, β is assumed to be constant.

An equation for R of this form was originally derived by Hitschfeld and Bordan [19], and can be considered as an extension of the backscattering method when attenuation is present. Although the estimates are unstable even with fairly modest errors in the radar constant or in the various power law relationships, if an independent measurement of rain rate were made at range r (using a rain gauge for example), the errors in the rain rate have been shown to be reduced substantially. Although such point measurements of rain rate are impractical for a spaceborne radar, several methods exist to provide independent estimates of the path attenuation or integrated rainfall rate (see Sections 5.3 and 5.4). Assuming, for example, that a measurement of path attenuation is made along the path from 0 to r_n, this can be used to obtain an independent estimate of the quantity $\exp(-0.2 \ln 10 \beta \int_0^{r_n} k(s) \, ds)$, which we denote by $p'(r_n)$. Letting $r = r_n$ in (5.5) gives a second estimate of p over the same interval. This second value of p can be made equal to $p'(r_n)$ either by modifying the radar constant C or the parameter α. In particular, if C is changed so that the right hand side of (5.5) is equal to $p'(r_n)$, then a substitution of the new value of C into (5.4) yields the modified rain rate estimate [10,11]:

$$\hat{R}_1(r) = aZ_m^b(r)\{0.46\beta[S(r_n)/(1 - p'(r_n)) - S(r)]\}^{-b/\beta}; \; r \leq r_n \qquad (5.7)$$

where

$$S(r) = \int_0^r \alpha Z_m^\beta \, ds; \quad r \leq r_n \qquad (5.8)$$

If α is adjusted rather than C, the following estimate is found:

$$\hat{R}_2(r) = aZ_m^\beta[1 - (1 - p'(r_n))S(r)/S(r_n)]^{-b/\beta}; \; r < r_n \qquad (5.9)$$

The rain rates given by Equations (5.7) and (5.9) are independent of biases in C and α, respectively.

In Figure 5.2 the curves labeled $\hat{R}_X$ and $\hat{R}_K$ represent the rain rates as derived from X-band (10 GHz) and Ka-band (35 GHz) airborne radar data, respectively.

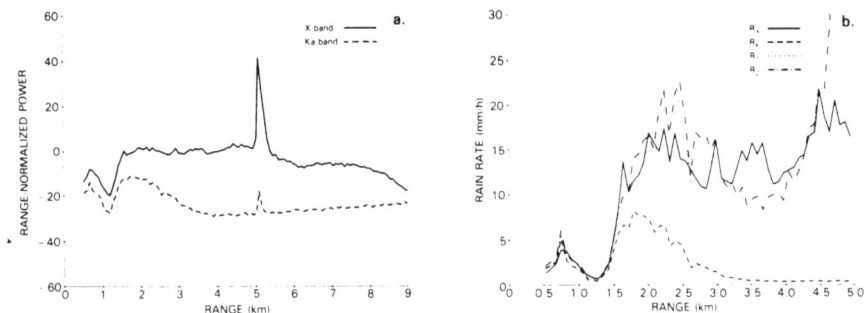

Figure 5.2 (a) Range-normalized return powers *versus* distance from an airborne radar for 10 GHz (solid line) and 35 GHz (dashed line) frequencies; (b) corresponding profiles of the rain rate as derived from the backscattering methods at $f = 10$ GHz (R_X) and at $f = 34.45$ GHz (R_K). R_1, R_2 represent estimates of rain rate using the 35 GHz radar with a path-attenuation correction.

These estimates were found by using the backscattering method without attenuation correction, Equation (5.1). At Ka-band, the effects of attenuation can be seen clearly. To attempt to correct the Ka-band estimate, (5.7) and (5.9) are employed where $p'(r_n)$ is obtained from a surface reference method at Ka-band (see Section 5.4). If the X-band derived rain rate is taken as the standard of comparison, we can see that attenuation correction can improve accuracy. Despite the improvements, however, instabilities can appear (e.g., Figure 5.2(b)) which are caused by low S/N. Other error sources in the method arise from inaccuracies in the estimation of $p'(r_n)$ and difficulties in apportioning the attenuation when the bright-band and other nonliquid hydrometeors are present.

Recently, an alternative single-wavelength algorithm has been proposed that employs a constraint [12]. By forming the ratio of return powers at adjacent range gates, taking the logarithm to the base 10 of both sides and substituting the relationships $Z_e = aR^b$ and $k = qR^p$, the following set of $(n - 1)$ equations is obtained for the rain rate, R, at each of the n range gates:

$$\log\{[P(r_{i+1})r_{i+1}^2]/[P(r_i)r_i^2]\} = b\,\log[R(r_{i+1})/R(r_i)]$$

$$- 0.2q \int_{r_i}^{r_{i+1}} R^p\,ds\ (i = 1, \ldots, n - 1) \quad (5.10)$$

Letting h be the range resolution, the last term is approximated by

$$\int_{r_i}^{r_{i+1}} R^p\,ds = 0.5h[R^p(r_i) + R^p(r_{i+1})] \quad (5.11)$$

If the parameters in the Z_e-R and k-R relationships are assumed to be known, then the above set constitutes $(n-1)$ equations for the n unknown rain rates. The final equation takes the form of an integral constraint:

$$\sum_{i=1}^{n} R(r_i)$$

which can be estimated from a microwave radiometer or a surface reference method. After linearization, the resulting set of equations can be solved [12]. Methods of this general type are well suited to a radar and microwave radiometer combination where the path-attenuation, deduced from a single or multichannel radiometer, is used to compensate for attenuation effects in the radar measurements [13–16]. We should also mention that methods similar to those described above have been developed independently for lidar sensing of aerosols and clouds [17].

5.3 DUAL-WAVELENGTH METHODS

To make the wavelength dependence of the radar equation explicit, we write:

$$P(\lambda_i,r) = C(\lambda_i)|K_w|^2 Z_e(\lambda_i,r) \exp(-0.46 \int_0^r k(\lambda_i,s)\ ds)/r^2 \qquad (5.12)$$

For a dual-wavelength radar with $\lambda_1 < \lambda_2$, an estimate of the differential attenuation in the range interval from r_k to r_j is [18–22]:

$$A = \int_{r_k}^{r_j} [k(\lambda_1,s) - k(\lambda_2,s)]\ ds = -5 \log(P_{12}/Z_{12}) \qquad (5.13)$$

where

$$P_{12} = [P(\lambda_1,r_j)P(\lambda_2,r_k)]/[P(\lambda_2,r_j)P(\lambda_1,r_k)] \qquad (5.14a)$$

$$Z_{12} = [Z_e(\lambda_1,r_j)Z_e(\lambda_2,r_k)]/[Z_e(\lambda_2,r_j)Z_e(\lambda_1,r_k)] \qquad (5.14b)$$

If Z_{12} is equal to unity, then the differential attenuation, A, can be expressed in terms of the measured quantity P_{12}. This condition is satisfied exactly if either the radar reflectivity factors are independent of wavelength at r_k and r_j, or if the reflectivity factor at each wavelength is uniform in range.

Correlation coefficients between the values of dBZ measured aloft and those computed from drop size distributions at the ground are found to decrease from about 0.8 for a 0.58 km range separation, to about 0.6 for a 3.6 km separation [23].

Although these results suggest the choice of a small value of $r_j - r_k$, as the interval decreases, the standard deviation in the return power, caused by finite sampling, ultimately becomes comparable to the attenuation. Thus, the optimum choice of the interval is dependent on the number of independent samples and the uniformity of the rain or cloud. Similar trade-offs are evident in the selection of wavelengths. While Z_e is independent of frequency in the Rayleigh regime, the attenuation is small. By increasing the separation between the two frequencies, the attenuation, and therefore its detectability, will increase, but usually at the expense of a loss in correlation between $Z_e(\lambda_1)$ and $Z_e(\lambda_2)$. Moreover, cloud liquid water can enhance the estimate of differential attenuation leading to overestimates in the rainfall rate [23]. Even at low frequencies, regions of strong non-Rayleigh scattering such as the bright-band or hail must be excluded.

Although it is sometimes stated that the dual-wavelength methods are independent of the radar constant, the statement is valid only if the return power can be represented by (5.12). If the partial beamfilling (or reflectivity gradients) vary within the interval over which the method is applied, the radar "constants" in general will not cancel by taking the ratio of powers [24,25]. A related problem arises when mismatches exist in the antenna patterns of the two frequencies, particularly when strong reflectivity gradients are present [26]. One of the original applications of the dual-wavelength method was in the detection of hail, where either the ratio of return powers at 10 cm to that at 3 cm [27] or the range derivative of that ratio [28] was used. Recently, Bringi *et al.*, [29] have employed a dual-wavelength ratio with attenuation correction and found it to be well correlated with dual-polarization indications of hail.

As noted in Section 5.2, when the rain or cloud is uniform in range, the specific attenuation can be deduced from a single-wavelength radar. For a dual wavelength radar, a two-parameter DSD can be estimated. Goldhirsh and Katz [30] have shown that this procedure can be used for both attenuating or nonattenuating wavelengths and dual-attenuating wavelengths. Moreover, for the latter combination, subject to the restrictions discussed above, the estimates are independent of the radar constant. To outline the method in the case when both wavelengths are attenuated, we let

$$T_i = [P(\lambda_i, r_j) r_j^2] / [P(\lambda_i, r_k) r_k^2] \tag{5.15}$$

Assuming that the drop size distribution, $N(D)$, is independent of range in the interval $[r_k, r_j]$, then

$$\log T_1 / \log T_2 = k_1 / k_2 =$$
$$= \int_D \sigma_t(\lambda_1, D) N(D) \, dD / \int_D \sigma_t(\lambda_2, D) N(D) \, dD \tag{5.16}$$

where σ_t is the extinction cross section. Expressing $N(D)$ as $N_0 D^m \exp(-\Lambda D)$ where m is specified, substituting this into (5.16) and noting that the N_0 factors cancel, the equation yields an estimate of Λ in terms of the measured quantities on the left-hand side. The parameter N_0 follows from the use of either T_1 or T_2:

$$N_0 = -5 \log T_i / [0.434(r_j - r_k) \int \sigma_t(\lambda_i, D) D^m e^{-\Lambda D} \, dD] \tag{5.17}$$

Algorithms for the attenuating (λ_1) and nonattenuating (λ_2)pair follow from the use of an analogous procedure: the ratio $k(\lambda_1)/Z_e(\lambda_2)$ is independent of N_0 and provides a solution to Λ. N_0 is then obtained by noting that the measured $Z_e(\lambda_2)$ can be expressed as a function of N_0 and Λ. Error analyses of these methods indicate that the number of independent samples that compose the estimate of the return power must be fairly large to yield reasonably accurate estimates of $N(D)$ [31–32]. This sensitivity to the sampling number arises because the estimates are formed by ratios of powers, which has the effect of amplifying the error variance.

In the method just outlined, the size distribution is assumed to be uniform over the interval $[r_k, r_j]$. In the case of attenuating and nonattenuating wavelengths, this restriction can be relaxed somewhat by writing the distribution in the form [33]:

$$N(D,r) = N_0(r) D^m \exp(-\Lambda D) \tag{5.18}$$

where m is specified and Λ is assumed to be range-independent. To determine $N_0(r)$ and Λ, using Equation (5.13) with $\log Z_{12} = 0$ (5.14b), then

$$5 \log P_{12} = -\int_{r_k}^{r_j} k(\lambda_1, s) \, ds$$

$$= -0.434 \int_{r_k}^{r_j} N_0(s) \, ds \int_D \sigma_t(\lambda_1, D) D^m \exp(-\Lambda D) \, dD \tag{5.19}$$

Moreover, from the definition of $Z_e(\lambda_2)$:

$$\int_{r_k}^{r_j} Z_e(\lambda_2, r) \, dr = H \int_{r_k}^{r_j} N_0(s) \, ds \int_D \sigma_b(\lambda_2, D) D^m \exp(-\Lambda D) \, dD \tag{5.20}$$

where

$$H = 10^6 \lambda_2^4 / |K_w|^2 \pi^5 \tag{5.21}$$

The ratio of these equations is independent of $N_0(r)$ and provides an estimate of Λ in terms of measured quantities. The coefficients $N_0(r_i)$ are then determined from the radar equation at λ_2:

$$N_0(r_i) = r_i^2 P(\lambda_2, r_i)/[C(\lambda_2)I(m\Lambda)H] \tag{5.22}$$

where

$$I(m\Lambda) = \int \sigma_b(\lambda_2, D)D^m e^{-\Lambda D} \, dD \tag{5.23}$$

Assuming a velocity-drop size relationship $v(D)$ (m/s), the rain rate (mm/h) estimate becomes:

$$R(r_i) = 0.6\pi N_0(r_i) \int_D D^{3+m} v(D)e^{-\Lambda D} \, dD \tag{5.24}$$

One other method that has recently been proposed begins with a distribution of sizes so that N_0 is fixed along the path but Λ is allowed to vary with range [34]. To give a simplified description, we assume that $k(\lambda_1) = C_k(\lambda_1) M_p$ and $Z_e(\lambda_2) = C_z(\lambda_2) M_q$, where M_x is the xth moment of DSD, and C_k and C_z are multiplicative constants. From a DSD with two free parameters (e.g., a gamma distribution with the unknowns Λ and N_0, and m specified), a k-Z_e relationship, $k(\lambda_1) = aZ_e(\lambda_2)^b$, can be expressed as a function N_0. Employing this relationship, the path-averaged estimate of N_0 is given by:

$$\hat{N}_0 = \left\{ \int k(\lambda_1, s) \, ds / [a' \int Z_e^b(\lambda_2, s) \, ds] \right\}^{1/(1-b)} \tag{5.25}$$

with

$$a = \frac{C_k(\lambda_1)\Gamma(m + p + 1)}{[C_z(\lambda_2)\Gamma(m + q + 1)]^b} N_0^{1-b} = a'N_0^{1-b}, \text{ and } b = \frac{m + p + 1}{m + q + 1}$$

$\int k(\lambda_1, s)ds$ is the path attenuation, which can be estimated either from the radar surface return or the microwave radiometric brightness temperature. The parameter a' is a function of frequency, temperature and the parameter m, but independent of Λ and N_0. In the case of Rayleigh scattering, $p = 3$, $q = 6$, and C_k and C_z are defined in Table 4.5.

Once $\hat{N}_0$ is found from this equation, $\Lambda(r)$ can be determined at each range gate by using the relationship between $Z(\lambda_2, r)$, N_0, and $\Lambda(r)$. Figure 5.3 shows a

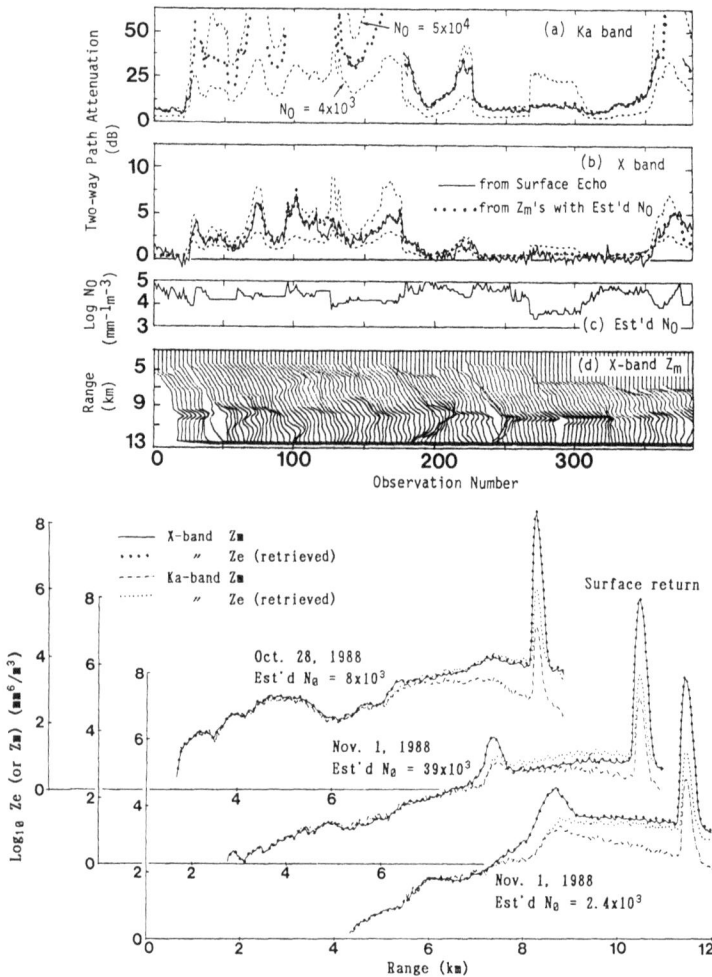

Figure 5.3 (a) Comparisons of the estimates of the two-way path attenuation: solid line represents the attenuation derived from the surface echo at Ka-band (35 GHz); the upper and lower dashed curves represent the attenuation based on $N_0 = 5 \times 10^4$ and $N_0 = 4 \times 10^3$ mm^{-1} m^{-3}, respectively; the dotted line represents the attenuation derived from the estimated N_0 values shown in c. (b) Same as (a) except for the X-band (10 GHz) radar. (c) Estimated values of N_0. (d) Range-profile of the measured X-band reflectivity factors where the range is measured from the aircraft. (e) Examples of the reconstruction of the Z_e profile using the estimated values of N_0 (after Kozu *et al.* [34]).

time plot of the estimated N_0 using data from an airborne dual-wavelength radar; also shown are the measured reflectivity factors at X-band where the range is measured from the airborne radar as well as examples of the reconstructed Z_e

profile based on estimated DSD. Although it is too early to assess this technique, the estimated values of N_0 are reasonable in that they generally fall within the range of commonly measured N_0 distributions and that shifts in the N_0 values appear to be correlated with changes in storm type [34].

A different approach to the dual-wavelength problem has been proposed by Fujita [35] for attenuating and nonattenuating wavelengths. By forming the ratio of powers at adjacent range bins and taking the logarithm of the result, then for n range gates and two wavelengths, a set of $2(n - 1)$ equations is obtained. Expressing Z_e and k in terms of R by the relationships $Z_e = aR^b$ and $k = \alpha R^\beta$, then the set of $2(n - 1)$ equations contains $(n + 4)$ unknowns: the n rain rates, one per range gate, the values of b at both wavelengths, and the values of α and β at the attenuating wavelength. Note that the equations are independent of the parameter a. Because the k-R relationship is relatively insensitive to changes in the DSD, the parameters α and β are assumed to be known. This leaves $(n + 2)$ unknowns in $2(n - 1)$ equations which for $n > 4$ can be solved in a least squares sense. Because the equations are nonlinear, the solution is obtained asymptotically by means of an iterative solution of the linearized equations. Although this procedure has similarities to the methods outlined in Section 5.2, in this case a type of dual-wavelength constraint is applied in a least mean squares sense over the interval. Other methods which use dual-wavelength constraints have been formulated [11,16].

5.4 SURFACE REFERENCE METHODS

A fixed target such as a corner reflector placed in the far field of the radar can provide an estimate of path attenuation if the ratio of backscattered powers is measured in both the presence and absence of rain [36–38]. For a spaceborne weather radar in the presence of rain along the beam, the scattering cross section of the surface, σ°, is generally unknown. In some cases, however, measurements at an adjacent rain-free area or a measurement made at the same location at times when rain is absent can serve as a reference [10,39–40]. We assume for simplicity that the returns from the surface, in the presence and absence of rain, can be written respectively as

$$P_s(r_s) = C_s \sigma^\circ \exp\left(-0.46 \int_0^{r_s} k \, ds\right) \tag{5.26}$$

$$\bar{P}_s(r_s) = C_s \bar{\sigma}^\circ \tag{5.27}$$

where C_s represents the radar and range dependencies for either a beam or pulse limited situation (see Section 2.2), and we assume that the measurements within and outside the rain are made at the same incidence angle and altitude. Assuming that $\sigma° = \bar{\sigma}°$, the path attenuation can be estimated from the equation

$$A(r_s) = \int_0^{r_s} k(s) \, ds = 5 \log[\bar{P}_s(r_s)/P_s(r_s)] \tag{5.28}$$

If the surface return exceeds both the rain return and the receiver noise power at the pulse volume centered at r_s, then the largest source of error arises from fluctuations in $(\sigma°/\bar{\sigma}°)$. Variations in $\sigma°$ for vegetation are given by Ulaby [41] for frequencies below 18 GHz: the standard deviation in $\sigma°$ decreases monotonically from a maximum value of about 6 dB at nadir to 20°; beyond 20°, the standard deviation is fairly constant with a value of about 2 dB at 17 GHz. Over the ocean, measurements of $\sigma°$ at $f = 13.9$ GHz have been plotted as a function of windspeed for various incidence angles [42]. For angles less than 20°, $\sigma°$ varies by less than 10 dB for windspeeds between one and 20 m/s; near 10° $\sigma°$ is approximately independent of windspeed. For incidence angles beyond about 30°, the sensitivity of $\sigma°$ to windspeed suggests that the method is unreliable for all but high values of attenuation unless the reference value $\sigma°$ is highly correlated with that measured in rain.

Apart from the change in $\sigma°$ with windspeed, another source of error arises from fluctuations in $\sigma°$ caused by the raindrops striking the surface. Both theoretical [43] and experimental studies [44–45] of this effect have been conducted. Although these studies undoubtedly will aid in the interpretation of the measurements, the functional relationship between the surface cross section and rain rate is not a simple one, because the former quantity also depends on wind speed, incidence angle, and frequency. If the correlation between $\sigma°$ and $\bar{\sigma}°$ is small, it is preferable to form the reference measurement from a spatial average of $\bar{\sigma}°$ (over a fixed angle and uniform surface conditions) to reduce sampling errors. For measurements over the ocean, Fujita et al. [40] have used the assumption that $\bar{\sigma}° = 10$ dB at nadir for $f = 35$ GHz. Comparisons between the backscattered power from the surface versus path attenuation have shown these quantities to be well correlated, especially at the higher values of attenuation (Figure 5.4). Over land, the reference measurement of $\sigma°$, taken over the same location as that of the rain at a prior or subsequent time, is preferable to a spatial average especially if the terrain is changing.

Dual-wavelength versions of this method have also been analyzed [46–48]. In one version of the method, the differential attenuation estimate takes the form

$$A_{12}(r_s) = \int_0^{r_s} [k(\lambda_1,s) - k(\lambda_2,s)] \, ds$$

$$= 5 \log \left[\frac{\bar{P}_s(\lambda_1,r_s)P_s(\lambda_2,r_s)}{P_s(\lambda_1,r_s)\bar{P}_s(\lambda_2,r_s)} \right] \tag{5.29}$$

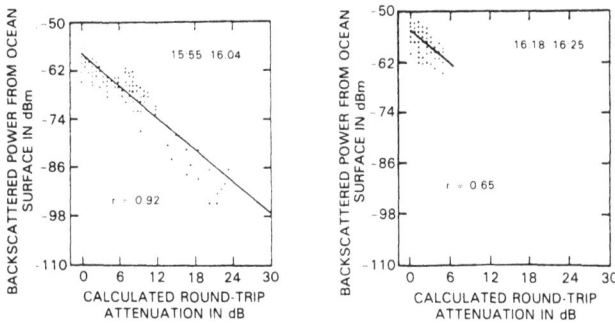

Figure 5.4 Backscattered powers from the ocean surface at nadir incidence *versus* the round-trip attenuation (from Fujita *et al.* [40]).

where $P_s(\lambda_i,r_s)$, $\bar{P}_s(\lambda_i,r_s)$ are the return powers from the surface at wavelength λ_i in the presence and absence of rain, respectively. The primary advantage of the dual-wavelength version of the method is that the variance of the attenuation estimate is reduced in proportion to the degree of correlation between $\sigma°$ at the two wavelengths.

Comparisons of Methods

Comparisons between rain rate estimates deduced from the dual-wavelength, the surface reference, and the backscattering methods are shown in Figure 5.5. These results were derived from a dual-wavelength airborne experiment for observations near nadir. In each plot, the rain rate is shown as a function of the observation number. A single observation represents a sample average of 128 waveforms over about a $\frac{1}{3}$ s interval. Figure 5.5(a) presents a comparison of rainfall rates as deduced from the backscattering methods applied to the 10 GHz data (Z_X-R) and to the 35 GHz data (Z_K-R). Even for these light rain rates, the attenuation at the higher frequency is strong enough to produce a negative bias in the Z_K-R estimates. In Figure 5.5(b), results of the surface reference (dual-wavelength) are compared with the Z_X-R results. The surface reference method, like the dual-wavelength method pictured in Figure 5.5(c), follows the Z_X-R results fairly well, but exhibits fine scale oscillations that are probably unrelated to the actual rain rate.

It is worth noting that the methods have different ranges of applicability, and that the dynamic range of each method is in general a function of frequency, incidence angle, gain and transmit power. Figure 5.6 shows a schematic of the dynamic range for several methods, using parameters for a two-frequency radar at 14 GHz and 24 GHz for the case of nadir viewing over an ocean background. The

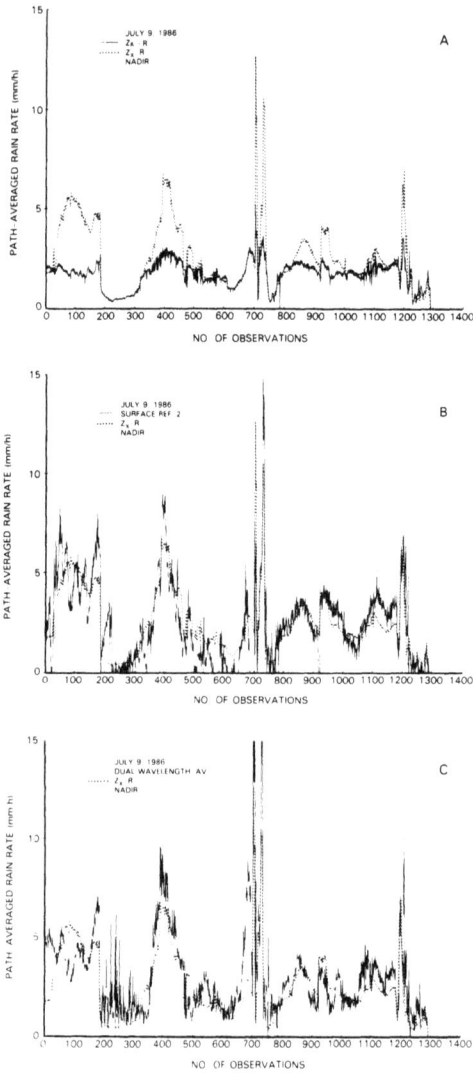

Figure 5.5 Comparisons of path-averaged rain rate estimates *versus* time (observation number) for (a) backscattering methods at $f = 10$ GHz (Z_X-R) and $f = 34.45$ GHz (Z_K-R); (b) surface reference method (dual-wavelength) and Z_X-R; (c) dual-wavelength method and Z_X-R (after Meneghini *et al.* [46]).

point of the figure is not the specific values, which depend on the radar design and storm model, but that the dynamic range corresponds with certain features of the signal. For example, the single-wavelength *surface reference method* (SRT) can

NEAR NADIR/OCEAN

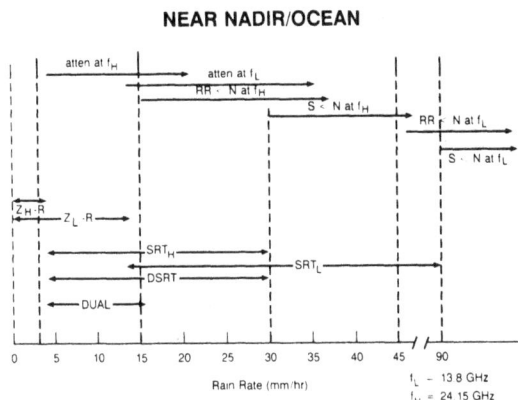

Figure 5.6 Approximate dynamic range for the Z-R, single (SRT) and dual-wavelength (DSRT) surface reference, and the dual-wavelength methods (DUAL) as correlated with the relative magnitudes of the rain (RR), surface (S) and receiver noise (N) for near-nadir incidence over the ocean. Subscripts L and H denote the lower and higher frequencies, respectively, where f_L = 13.8 GHz, f_H = 24.15 GHz.

be applied only if the surface return exceeds the noise level and if the rain attenuation is larger than the fluctuations in σ°. The dynamic ranges of the single and dual-wavelength SRT are reduced substantially over land or at angles far from nadir (see Section 2.2). On the other hand, the dual wavelength technique is independent of the surface and relatively insensitive to incidence angle, but can be applied only over paths where the rain signals at both frequencies exceed the system noise level and where attenuation at the higher frequency is larger than the standard deviation in the estimate. Conversely, for the backscattering method, a necessary condition for an accurate estimate of rain rate is that the path attenuation be small. The primary error sources in the backscattering, dual-wavelength, and surface reference method are virtually independent, so that in cases where all three methods give the same rain rate, some confidence in the results is warranted [46]. The methods are also complementary in the sense that the attenuation methods are useful at the higher rain rates, and the backscattering method is useful at the low rain rates.

5.5 OTHER ATTENUATION METHODS

Mirror-Image (MI)

The mirror-image return (see Chapter 2) has been considered from the standpoints of both attenuation and velocity estimation. The ratio of the MI to the direct rain return is a function of the backscattering cross section, the Fresnel reflectivity of the surface, and the four-way path attenuation from the surface to the height at which the direct and MI returns are measured. Several methods have been proposed to remove the dependence on the unknown backscattering cross section either by forming the ratio of the MI return to the direct return over the same volume of rain or by expressing $\sigma°$ in terms of the measured surface return power [49–50]. Although some limited experimental results have been analyzed, it is premature to judge whether the methods are feasible. Optimum conditions for its application are at near nadir angles over the ocean employing a narrow beam antenna. Error sources in the method include those given below:

(1) Mismatches between the direct and mirror-image rain volumes are sources of error. The problem is less severe for typical spaceborne geometries than for airborne geometries because of the smaller beamwidths and much greater altitudes used in the former. For off-nadir angles and for rough surfaces, the mismatches increase.

(2) Because the MI return is proportional to the four-way rather than the two-way path attenuation, the sensitivity at light rain rates is improved relative to other attenuation methods. By the same token, as the rain intensity increases, the MI will be masked by the system noise because of excessive attenuation.

(3) Due to the smaller values of $\sigma°$ and surface reflectivity, the strength of the MI return over land at near-nadir incidence should typically be over 10 dB smaller than over ocean.

(4) At present, the algorithms for rain attenuation are based on a nonrigorous derivation of the MI return. Further theoretical and experimental work is needed to test the validity of the theory.

Stereoscopic Methods

Additional information on the scattering medium can be inferred if the same volume of hydrometeors is viewed from more than one direction. This idea has been used to derive two-dimensional wind fields from ground-based and airborne radars [51]. From an airborne or spaceborne radar, the two viewing angles can be obtained either by a conically scanning antenna or by a dual antenna system, at least one of which scans in a plane tilted relative to the cross track plane. Although the primary use of this geometry has been for doppler applications, the potential of stereo attenuation measurements also has been explored [52,53]. The methods derive from the fact that if the same volume of rain can be viewed at two angles, then the ratio of powers is independent of the radar reflectivity of the volume and proportional only to the attenuations along the two paths. We note that if the energy scattered from the surface is idealized as being directed along the specular direction, then the MI return also may furnish a second viewing angle of each rain volume. Use of the MI return for deriving two-dimensional wind fields in this manner has been investigated by Atlas and Matejka [54].

5.6 CLIMATOLOGICAL METHODS

One of the basic quantities of interest in climatology and hydrology is the volumetric rainfall rate (volume/time) or rainfall amount over an area (volume). That such measurements can be made with reasonable accuracy without measuring the rain rate at each resolution element has been demonstrated by the methods developed for the visible and infrared radiometers. Accounts of these methods are given by Barrett and Martin [55]. One of the difficulties in such methods is delineation of the raining areas by means of the time histories of cloud top and cloud integrated brightness temperatures. The advantage of radar is that the rain can be sensed directly, eliminating the need for proxy variables. Compared to the sensors on geostationary satellites, however, the temporal resolution of the radar (confined to an orbiting platform) will be poor.

Byers [56] appears to be the first to have established a relationship between volumetric rainfall and the size of a convective storm. From convective storms in Kansas, Crane [57] and Crane and Hardy [58] noted that by counting the number and spacings of significant convective cells, the total rainfall over an area could be estimated to within a factor of 2 or less. Donneaud *et al.* [59] established a relationship between the rain volume, V, and a quantity called the *area-time integral* or ATI (originally termed the *integrated rainfall coverage*), where

$$V = \int_0^T \int_A R \, dA \, dt \qquad (5.30)$$

and

$$\text{ATI} = \int_0^T \int_{A'} dA \, dt \qquad (5.31)$$

where R is the rainfall rate in mm/h and the area of integration, A, is taken to be equal to the total raining area. The area integration, A', over which the ATI is computed, is taken over all regions for which the radar reflectivity (or rainfall rate) exceeds some threshold. Donneaud *et al.* [60] have taken this to be 25 dBZ.

Donneaud *et al.* [59] have also found that for areas on the order of 10^4 km^2 and for total time of observation of 12 hours with a sampling rate of once per hour, the scatter plots of V *versus* ATI on a log-log scale yielded a correlation coefficient of 0.955 for a set of 44 measurements. Although these results are not directly applicable to a single orbiting platform where the revisit times are only once or twice per day, the potential exists for extending such methods to an orbiting radar and a geostationary infrared sensor. One possibility is using the radar as a type of calibration to search for combinations of measurable quantities from the infrared sensor that remain relatively constant along the contours of uniform dBZ or rainfall rate. This calibration could then be extended to observations when the radar is not present.

A closely related issue is the relationship between the area within some dBZ or rain rate threshold and the area average rainfall rate at an instant of time. Figure 5.7 shows a sequence of scatter plots of the area average rainfall rate (mm/hr) *versus* the fractional rain area (ratio of the area within which the rain rate exceeds a given threshold level to the total area) for five threshold levels. As seen from the figure, the threshold value of 5 mm/h minimizes the amount of scatter and yields a correlation coefficient of approximately 0.98 [61]. These results were obtained using the GATE data set over a 280 km × 280 km region. Chiu [61] notes that the 5 mm/h threshold effectively defines the area occupied by the convective precipitation (omitting the stratiform portion of the storm cells), so that it is similar to the cloud indexing methods that have been used in infrared sensing [62]. Recently, extensions of the method have been made by Rosenfeld *et al.* [63] and Atlas *et al.* [64], showing that further improvements in the estimation of the areal average rainfall rates can be made by using both area and height thresholds to characterize the raining region. These last two papers also discuss the theoretical basis for such

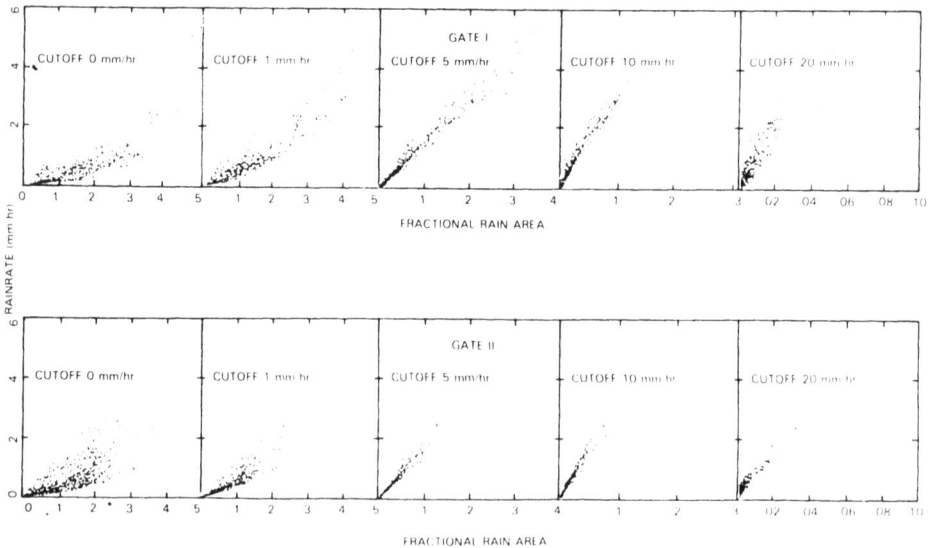

Figure 5.7 Area average rain rate *versus* the fractional rain area, as defined by five rain rate threshold levels for the GATE phases I and II (from Chiu [61]).

methods, linking their effectiveness to the existence of a consistent probability density function of rain rate for each climatic region.

Some of the potential advantages of these and other climatological methods follow:

(1) The threshold area (and height) characterization of the rain is highly correlated with area rainfall if the area over which the fitting is applied is greater than about 100 km on a side. Measurements from space of the former quantities appear to impose fewer demands on the measuring system and the data processing.

(2) If threshold area (and height) is sufficient to characterize the volumetric rainfall rate, then adaptive scanning methods (see Section 3.3) could be used in an optimum manner to conserve spacecraft power or increase the swath. For sensors such as TRMM, the narrow-swath radar data can serve as a calibration that may be applied to the large-swath sensors such as microwave and infrared radiometers.

(3) Further developments of these and other methods should help identify sampling strategies for the statistical characterization of rainfall from orbiting platforms.

Possible drawbacks are that methods of this type do not seem applicable to purely stratiform rain. Moreover, establishing the relationships between area thresholds and area rainfall rates in various climatological regions requires an initial calibration using either traditional radar methods or the gauge-radar distribution matching procedure (see Section 5.1).

5.7 POLARIMETRY METHODS

One of the most significant advances in radar meteorology since the mid-1970s has been the introduction and testing of a number of polarimetry methods. Falling raindrops tend to assume an oblate spheroidal shape, with the symmetry axis along the vertical and an eccentricity proportional to the equivolume drop diameter [65–66]. Nonliquid and mixed phase hydrometeors, on the other hand, occur in a variety of shapes and orientations [67–68]. These and other observations have promoted the development of two kinds of polarimetry methods. If the raindrops have a common alignment and a size-dependent shape, estimates of the drop size distribution, as well as rainfall rate and liquid water content, can be made. The second type of method has focused on detection and discrimination: oriented particles such as raindrops should be distinguishable from randomly oriented, irregularly shaped, or spherically shaped hydrometeors by their polarization signature.

The introduction of the *differential reflectivity ratio*, ZDR, [69–70] provided the impetus for a large number of investigations, including studies of scattering from nonspherical scatterers [71–72], particle shape and orientation [66,73–75], and discrimination of hail from rain [29,76]. A number of theoretical and experimental studies have addressed the potential of such methods to estimate the drop size distribution, rain rate and attenuation [77–79]. Improvements in estimation from the addition of phase information or dual-wavelength data also has been addressed [29,74,79]. Relationships between the ZDR, the *circular depolarization ratio*, CDR, and the *linear depolarization ratio*, LDR, are given by Stapor and Pratt [80] and Hendry *et al.* [81]. In its simplest form, the ZDR technique for the estimation of the drop size distribution rests on several assumptions: (1) the raindrops are oblate spheroids with symmetry axes (canting angles) that are highly concentrated about the vertical; (2) a one-to-one correspondence exists between the shape and size of the raindrop—specifically, the relationship $r = 1.03 - 0.062 D$ is often used [65], where r is the ratio of the minor to the major axis of the oblate spheroid and D is the diameter (mm) of an equivolume sphere; (3) the drop size distribution is in the form of a gamma distribution, $N(D) = N_0 D^m \exp(-\Lambda D)$, where m and the minimum and maximum equivolume diameters are assumed. Because ZDR is given by the ratio of the two orthogonal copolarized radar reflectivity factors, it is independent of N_0, and therefore can be used to estimate Λ, or, equivalently the median drop diameter, D_0. N_0 is then estimated from either of the two linear co-polarized returns.

Calculations of ZDR and LDR are shown in Figure 5.8 as functions of D_0 for five values of the parameter m [29]. These calculations were performed for horizon viewing at S-band (3 GHz) for a Gaussian distribution of canting angles with a mean of $2°$ relative to the vertical and a standard deviation of $10°$. For horizon viewing, the ZDR values in rain are maximized when the polarization vectors are aligned along the major and minor axes of the vertically oriented raindrop. As nadir incidence is approached, ZDR (Z_{HH}/Z_{VV}) will approach to unity. Because most proposed space-based weather radars have maximum off-nadir angles below about $45°$, the dynamic range of ZDR (as well as CDR and LDR) will be reduced.

To date, most polarimetric meteorological radars have been operated below about 6 GHz. As the frequency is increased, propagation effects to and from the scattering volume become a source of error. These error sources include differential attenuation in the two polarization states, as well as depolarization of the incident and scattered fields in the forward scattering directions. Because the received field is a function not only of the backscattering properties of the hydrometeors in the range gate of interest, but also of propagation effects to and from the range gate, the estimation problem is analogous to a single attenuating wavelength radar where the number of unknowns exceeds the number of measurable quantities. Moreover, as the frequency increases, so does the number of cases for which the Rayleigh-Gans approximation is not valid. When this occurs, the high

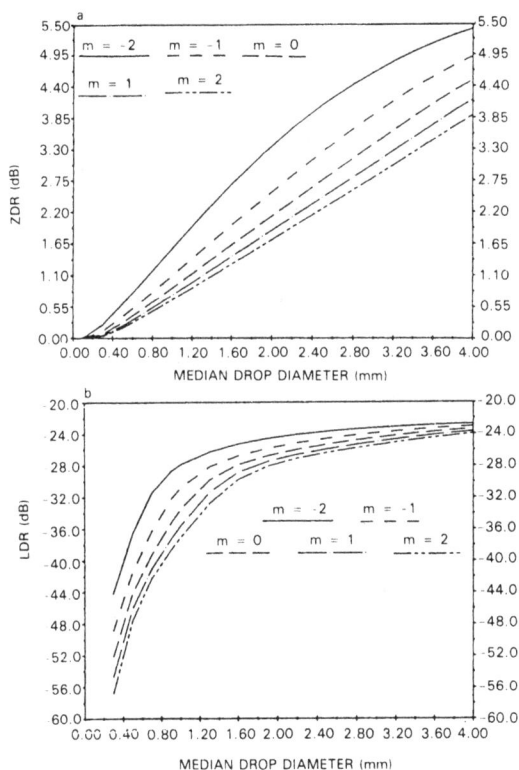

Figure 5.8 (a) ZDR *versus* median drop diameter for five values of the parameter *m* in the gamma drop size distribution; (b) same as (a), but for LDR (from Bringi *et al.* [70]).

correlations between the polarimetry measurements on one hand and shape and orientation effects on the other are no longer guaranteed. Calculations of ZDR and CDR *versus* the median volume drop diameter, D_0, ($m = 0$) for a number of radar frequencies has been presented by Holt [82]. Beyond about 20 GHz, the ZDR is insensitive to changes in the D_0 for all but light rain rates. The sensitivity of CDR to D_0 is somewhat better than ZDR at the higher frequencies, although the antenna isolation must be quite good to measure the smaller D_0 values. Nakamura *et al.* [83] found significant degradation in the ZDR method at 35 GHz relative to that at 10 GHz. To obtain an accurate estimate of ZDR, successive pairs of Z_{HH} and Z_{VV} measurements should be highly correlated but with a sufficient time lag between each Z_{HH}, Z_{VV} pair to achieve statistical independence [84]. From space, this can be accomplished by the simultaneous transmission of two orthogonally polarized signals, but at the expense of greater radar complexity and a loss in the SNR per pulse. Like the standard backscattering relationship, most quantitative polarimetry methods depend on a knowledge of the radar constant. Relative errors in the constants can be removed by noting that ZDR in rain should be near unity at nadir incidence.

Many polarimetry studies have been concerned with distinguishing various types of precipitation. Barge [85] noted that at a frequency of 3 GHz the probability of hail is zero for reflectivity factors less than 30 dBZ, and is always present for values exceeding 50 dBZ. At intermediate reflectivities, the presence of hail was well correlated with the degree of depolarization. More recently, the use of CDR and related quantities have been used to separate shape and orientation effects in snow, rain and in the melting level [82]. Measurements of ZDR in hail, where ZDR is near zero, have shown the signature to be clearly distinguishable from rain, where typical ZDR values occur in the range from 0.5 to 4 dB. The small values of ZDR in hail appear to be a consequence of the mean shape tending to be more spherical and more randomly oriented than raindrops [29]. Measurements of Z_{HH} and ZDR within and below the melting layer have shown that the ZDR has a well defined peak that occurs several hundred meters below that of Z_{HH}. The explanation for this offset is that the partially melted drops tend to assume an oblate shape only after the hydrometeor is almost entirely liquid water [29].

REFERENCES

[1]   Stout, G.E., and E.A. Mueller, 1968: Survey of relationships between rainfall rate and radar reflectivity in the measurement of precipitation. *J. Appl. Meteor.*, **7**, 465–474.

[2]   Wilson, J.W., and E.A. Brandes, 1979: Radar measurement of rainfall—a summary. *Bull. Amer. Meteor. Soc.*, **60**, 1048–1058.

[3]   Joss, J.K., and E.G. Gori, 1978: Shapes of raindrop size distributions. *J. Appl. Meteor.*, **17**, 1054–1061.

[4]   Calheiros, R.V., and I.I. Zawadzki, 1987: Reflectivity-rain rate relationships for radar hydrology in Brazil. *J. Clim. Appl. Meteor.*, **26**, 118–132.

[5]   Joss, J., and A. Waldvogel, 1989: Precipitation measurement and hydrology, Ch. 29a in *Radar in Meteorology*, D. Atlas, ed., Amer. Meteor. Soc., Boston.

[6]   Atlas, D., D.A. Short, and D. Rosenfeld, 1989: Climatologically tuned reflectivity-rain rate relations. *J. Appl. Meteor.*, **28**.

[7]   Atlas, D., and H.C. Banks, 1951: The interpretation of microwave reflections from rainfall. *J. Meteor.*, **8**, 271–282.

[8]   Goldhirsh, J., and E.J. Walsh, 1982: Rain measurements from space using a modified Seasat-Type altimeter. *IEEE Trans. Antennas and Propag.*, **AP-30**, 726–733.

[9]   Hitschfeld, W., and J. Bordan, 1954: Errors inherent in the radar measurement of rainfall at attenuating wavelengths. *J. Meteor.*, **11**, 58–67.

[10]   Meneghini, R., J. Eckerman, and D. Atlas, 1983: Determination of rain rate from spaceborne radar using measurements of total attenuation. *IEEE Trans. Geosci. and Remote Sensing*, **GE-21**, 34–43.

[11]   Meneghini, R., and K. Nakamura, 1989: Experimental tests of range-profiling algorithms. *Proc. 24th Conf. on Radar Meteor.*, Tallahassee, FL, March 27–31, 601–604.

[12] Fujita, M., 1989: An approach for rain rate profiling with a rain-attenuating-frequency radar under a constraint of path-integrated rain rate. *Proc. IGARRS*, 1491–1494.

[13] Lu, D., and L. Hai, 1980: Comparisons of radar and microwave radiometer in precipitation measurements and their combined use. *Acta Atmospherica Sinics*, 1.

[14] Hai, L., X. Miaoxin, W. Chong, and H. Yookui, 1985: Ground-based remote sensing of LCW in cloud and rainfall by a combined dual-wavelength radar-radiometer system. *Adv. Atmos. Sciences*, **2**, 93–103.

[15] Weinman, J.A., C.D. Kummerow, and C.S. Atwater, 1988: An algorithm to derive precipitation profiles from a downward viewing radar and a multifrequency passive radiometer. *IGARSS '88*, September, 13–16.

[16] Weinman, J.A., R. Meneghini, and K. Nakamura, 1989: Comparison of rainfall profiles retrieved from dual-frequency radar and from combined radar and passive microwave radiometric measurements. *Fourth Conf. on Satellite Meteor. and Oceanography*, May 16–19, San Diego, CA, 27–30.

[17] Reagan, J.A., M.P. McCormick, and J.D. Spinhirne, 1989: Lidar sensing of aerosols and clouds in the troposphere and stratosphere. *Proc. IEEE*, **77**, 433–448.

[18] Atlas, D., 1954: The estimation of cloud parameters by radar. *J. Meteor.*, **11**, 309–317.

[19] Eccles, P.J., and E.A. Mueller, 1971: X-band attenuation and liquid water content estimation by a dual-wavelength radar. *J. Appl. Meteor.*, **10**, 1252–1259.

[20] Eccles, P.J., 1979: Comparison of remote measurements by single- and dual-wavelength meteorological radars. *IEEE Trans. Geosci. and Remote Sensing*, **GE-17**, 205–218.

[21] Kostarev, V.V., and A.A. Chernikov, 1969: The adjustment of radar estimates of rainfall with radar attenuation data. Preprints *13th AMS Conf. on Radar Meteor.*, Amer. Meteor. Soc., Boston, pp. 396–399.

[22] Berjulev, G.P., and V.V. Kostarev, 1974: Dual wavelength radar measurements of 8 mm radiowave attenuation by atmospheric precipitation and clouds. *J. Rech. Atmos.*, **8**, 358–363.

[23] Joss, J., R. Cavalli, and R.K. Crane, 1974: Good agreement between theory and experiment for attenuation data. *J. Rech. Atmos.*, **8**, 299–318.

[24] Nakamura, K., 1989: A comparison of the rain retrievals by backscattering measurement and attenuation measurement. *Proc. 24th Conf. on Radar Meteor.*, Tallahassee, FL, March 27–31, 601–604.

[25] Amayenc, P., M. Marzoug, and J. Testud, 1989: Nonuniform beam filling effects in measurements of rainfall rate from a spaceborne radar. *Proc. 24th Conf. on Radar Meteor.*, Tallahassee, FL, March 27–31, 569–572.

[26] Rinehart, R.E., and J.D. Tuttle, 1982: Dual-wavelength processing—the effects of mismatched beam patterns. *Proc. URSI Commission F*, Multiple Parameter Measurements of Precipitation, Bournemouth, UK, August, 83–88.

[27] Jameson, A.R., and R.C. Srivastava, 1978: Dual-wavelength Doppler radar observations of hail at vertical incidence. *J. Appl. Meteor.*, **17**, 1694–1703.

[28] Eccles, P.J., and D. Atlas, 1973: A dual-wavelength radar hail detector. *J. Appl. Meteor.*, **12**, 847–856.

[29] Bringi, V.N., J. Vivekanandan, and J.D. Tuttle, 1986a: Multiparameter radar measurements in Colorado convective storms. Part II: Hail detection studies. *J. Atmos. Sci.*, **43**, 2564–2577.

[30] Goldhirsh, J., and I. Katz, 1974: Estimation of raindrop size distribution using multiple wavelength radar systems. Radio Sci., **9**, 439–446.

[31] Goldhirsh, J., 1975: Improved error analysis in estimation of raindrop spectra, rain rate, liquid water content using multiple wavelength radars. *IEEE Trans. Ant. and Propag.*, **AP-24**, 718–720.

[32] Stogyrn, A., 1975: Error Analysis of the Goldhirsh-Katz method of rainfall determination by the use of a two frequency radar. Report No. 1833TR-4, Aerojet Electrosystems Co., Asusa, CA.

[33]   Atlas, D., C.W. Ulbrich, and R. Meneghini, 1982: The multi-parameter remote measurement of rainfall. NASA Tech. Memo. 83971, July, 128 pp.

[34]   Kozu, T., R. Meneghini, W. Boncyk, K. Nakamura, and T.T. Wilheit, 1989: Airborne radar and radiometer experiment for quantitative remote measurements of rain. *Proc. IGARRS '89.*

[35]   Fujita, M., 1983: An algorithm for estimating rain rate by a dual-frequency radar. *Radio Sci.,* **18,** 697–708.

[36]   Collis, R.T.H., 1961: Digital processing of weather radar data. *Proc. 9th Wea. Radar Conf.,* Amer. Meteor. Soc., Boston, MA, 371.

[37]   Harrold, T.W., 1967: The attenuation of 8.6 mm wavelength radiation in rain. *Proc. Inst. Elec. Eng.* London, **114,** 201–203.

[38]   Atlas, D. and C.W. Ulbrich, 1977: Path- and area-integrated rainfall measurement by microwave attenuation in the 1-3 cm band. *J. Appl. Meteor.,* **16,** 1322–1331.

[39]   Eckerman, J., R. Meneghini, and D. Atlas, 1978: Average Rainfall Determination from a Scanning Beam Spaceborne Radar. NASA Tech. Memo. 79664, 35 pp. and appendices.

[40]   Fujita, M., K. Okamoto, S. Yoshikado, and K. Nakamura, 1985: Inference of rain rate profile and path-integrated rain rate by an airborne microwave scatterometer. *Radio Sci.,* **20,** 631–642.

[41]   Ulaby, F.T., 1980: Vegetation clutter model. *IEEE Trans. Ant. and Propag.,* **AP-28,** 538–545.

[42]   Jones, W.L., L.C. Schroeder, and J.L. Mitchell, 1977: Aircraft measurements of the microwave scattering signature of the ocean. *IEEE Trans. Ant. and Propag.,* **AP-25,** 52–61.

[43]   Manton, M.J., 1973: On the attenuation of sea waves by rain. *Geophys. Fluid Dynamics,* **5,** 249–260.

[44]   Moore, R.K., Y.S. Yu, A.K. Fung, D. Kaneko, G.J. Dome, and R.E. Werp, 1979: Preliminary study of rain effects on radar scattering from water surfaces. *IEEE J. Oceanic Eng.,* **OE-4,** 31–32.

[45]   Bliven, L.F., and G. Norcross, 1988: Effect of rainfall on scatterometer derived wind speeds. *Digest IGARSS 88,* WEA10-7.

[46]   Meneghini, R., K. Nakamura, C.W. Ulbrich, and D. Atlas, 1989: Experimental tests of methods for the measurement of rainfall rate using an airborne dual-wavelength radar. *J. Atmos. and Oceanic Tech.,* **6,** 637–651.

[47]   Meneghini, R., J.A. Jones, and L.H. Gesell, 1987: Analysis of a dual-wavelength surface reference radar technique. *IEEE Trans. Geosci. and Remote Sens.,* **GE-25,** 456–471.

[48]   Moore, R.K., 1981: Use of combined radar and radiometer systems in space for precipitation measurement—some ideas. In *Precipitation Measurements from Space:* Workshop Rept., Atlas, D., and Thiele, O.W., eds. NASA Goddard, Greenbelt, MD, pp D301–325.

[49]   Meneghini, R., and D. Atlas, 1986: Simultaneous ocean cross section and rainfall measurements from space with a nadir-looking radar. *J. Atmos. and Oceanic Technol.,* **3,** 400–413.

[50]   Meneghini, R., and K. Nakamura, 1988: Some characteristics of the mirror-image return in rain. *Proc. Intl. Symp. on Tropical Precip. Meas.,* Tokyo. October, 28–30, 235–242.

[51]   Hildebrand, P.H., and R.K. Moore, 1989: Meteorological radar observations from mobile platforms. Chapter in *Radar in Meteorology,* D. Atlas, ed., Amer. Meteor. Soc., Boston.

[52]   Weinman, J.A., 1984: Tomographic lidar to measure the extinction coefficients of atmospheric aerosols. *Appl. Optics,* **23,** 3882–3888.

[53]   Testud, J., and P. Amayenc, 1988: Stereoradar meteorology: A promising technique to observe precipitation from a mobile platform. *J. Atmos. Ocean. Tech.,* **6.**

[54]   Atlas, D., and T.J. Matejka, 1985: Airborne doppler radar velocity measurements of precipitation seen in ocean surface reflection. *J. Geophys. Res.,* **90,** 5820–5828.

[55]   Barrett, E.C., and D.W. Martin, 1981: *The Use of Satellite Data in Rainfall Monitoring,* Academic Press, London, 340 pp.

[56]   Byers, H.R., 1948: The use of radar in determining the amount of rain falling over a small area. *Eos Trans. AGU,* **29,** 187–196.

[57]   Crane, R.K., 1981: Sampling problems: The small scale structure of precipitation. In *Precipitation Measurements from Space:* Workshop Rept., Atlas, D., and O.W. Thiele, eds., NASA/GSFC, Greenbelt, MD, pp. D-41–D-49.

[58]   Crane, R.K., and K.R. Hardy, 1981: The HIPLEX program in Colby-Goodland Kansas: 1976–1980, Rep. P-1552-f, 144 pp., Environmental Research and Technology, Inc., Concord, MA.

[59]   Doneaud, A.A., P.L. Smith, A.S. Dennis, and S. Sengupta, 1981: A simple method for estimating convective rain volume over an area. *Water Resources Research,* **17,** 1676–1682.

[60]   Doneaud, A.A., S. Ionescu-Niscov, D.L. Priegnitz, and P.L. Smith, 1984: The area-time integral as an indicator for convective rain volumes. *J. Clim. Appl. Meteor.,* **23,** 555–561.

[61]   Chiu, L.S., 1988: Rain estimation from satellites: Area rainfall-rain area relation. *Third Conf. Satellite Meteor. and Oceanog.,* February 1–5, Amer. Meteor. Soc., Anaheim, CA, 363–368.

[62]   Arkin, P.A., 1979: The relationship between fractional coverage of high cloud and rainfall accumulations during GATE over the B-scale array. *Mon. Wea. Rev.* **107,** 1382–1387.

[63]   Rosenfeld, D., D. Atlas, and D.A. Short, 1988: The estimation of convective rainfall by area integrals, part II: The height area rainfall threshold (HART) method. *Conf. on Mesoscale Precip.,* MIT, Cambridge, MA, September 13–17.

[64]   Atlas, D., D. Rosenfeld, and D.A. Short, 1988: The estimation of convective rainfall by area integrals. Part I: The theoretical and empirical basis. *Conf. on Mesoscale Precip.,* MIT, Cambridge, MA, September 13–17.

[65]   Pruppacher, H.R., and K.V. Beard, 1970: A wind tunnel investigation of the internal circulation and shape of water drops falling at terminal velocity in air. *Quart. J. Royal Meteor. Soc.,* **96,** 247–256.

[66]   Jameson, A.R., and K.V. Beard, 1982: Raindrop axial ratios. *J. Appl. Meteor.,* **21,** 257–259.

[67]   Mon, J.P., 1982: Backward and forward scattering of microwaves by ice particles: A review. *Radio Sci.,* **17,** 953–971.

[68]   Seliga, T.A., and V.N. Bringi, 1976: Potential use of radar differential reflectivity measurements at orthogonal polarizations for measuring precipitation. *J. Appl. Meteor.,* **15,** 69–76.

[69]   Seliga, T.A., and V.N. Bringi, 1978: Differential reflectivity and differential phase shift: Applications in radar meteorology., *Radio Sci.,* **13,** 271–275.

[70]   Bringi, V.N., R.M. Rasmussen, and J. Vivekanandan, 1986 : Multiparameter radar measurements in Colorado convective storms. Part I: Graupel melting studies. *J. Atmos. Sci.,* **43,** 2545–2563.

[71]   Holt, A.R., 1982: The scattering of electromagnetic waves by single hydrometeors, *Radio Sci.,* **17,** 929–945.

[72]   Oguchi, T., 1981: Scattering from hydrometeors: A survey. *Radio Sci.,* **16,** 691–730.

[73]   Jameson, A.R., 1983: Microphysical interpretation of multi-parameter radar measurements in rain. Part I: Interpretation of polarization measurements and estimation of raindrop shapes. *J. Atmos. Sci.,* **40,** 1792–1802.

[74]   Jameson, A.R., 1985: Microphysical interpretation of multi-parameter radar measurements in rain. Part II: Estimation of raindrop distribution parameters by combined dual-wavelength and polarization measurements. *J. Atmos. Sci.,* **40,** 1803–1813.

[75]   Goddard, J.W.F., S.M. Cherry, and V.N. Bringi, 1982: Comparison of dual-polarization radar measurements of rain with ground-based disdrometer measurements. *J. Appl. Meteor.,* **21,** 252–256.

[76]   Hall, M.P.M., S.M. Cherry, and J.W.F. Goddard, 1980: Use of dual-polarization radar to measure rainfall rates and distinguish rain from ice particles. *IEEE Intern. Radar Conf.,* Washington, D.C.

[77]   Goddard, J.W.F., and S.W. Cherry, 1984: The ability of dual-polarization radar (copolar linear) to predict rainfall rate and microwave attenuation. *Radio Sci.,* **19,** 201–208.

[78]   Bringi, V.N., T.A. Seliga, and E.A. Mueller, 1982: First comparisons of rainfall rates derived from radar differ from radar differential reflectivity and disdrometer measurements. *IEEE Trans. Geosci. and Remote Sensing,* **GE-20,** 201–204.

[79]   Sachidananda, M., and D.S. Zrnic, 1986: Differential propagation phase shift and rainfall rate estimation. *Radio Sci.,* **21,** 235–247.

[80]   Stapor, D.P., and A. Pratt, 1984: A generalized analysis of dual-polarization radar measurements of rain. *Radio Sci.,* **19,** 90–98.

[81]   Hendry, A., Y.M.M. Antar, and G.C. McCormick, 1987: On the relationship between the degree of preferred orientation in precipitation and dual-polarization radar echo characteristics. *Radar Sci.,* **22,** 37–50.

[82]   Holt, A.R., 1984: Some factors affecting the remote sensing of rain by polarization diversity radar in the 3- to 35-GHz frequency range. *Radio Sci.,* **19,** 1399–1412.

[83]   Nakamura, K., H. Inomata, and J. Awaka, 1984: Polarization diversity measurements by *X*- and *Ka*-band radar. Preprints *22 Conf. on Radar Meteor.,* Amer. Meteor. Soc., Boston, MA, pp. 374–377.

[84]   Bringi, V.N., T.A. Seliga, and S.M. Cherry, 1983: Statistical properties of the dual-polarization differential reflectivity (ZDR) radar signal. *IEEE Trans. Geosci. and Remote Sensing,* **GE-21,** 215–220.

[85]   Barge, B.L., 1974: Polarization measurements of precipitation backscatter in Alberta. *J. Rech. Atmos.,* **8,** 163–173.

INDEX

THE AUTHORS

Robert Meneghini holds an MSEE from the University of Michigan. He spent two years in research in the Construction Engineering Research Lab at the University of Illinois, and the Space Physics Lab at the University of Michigan. He has been an Electrical Engineer with NASA/Goddard since 1974 and is a Member of the IEEE.

Toshiaki Kozu earned his MSEE from Kyoto University, Japan. In 1977, he became a Research Scientist for the Radio Research Lab (now the Communications Research Lab) of the Ministry of Posts and Telecommunications where he is now a Senior Research Scientist. He is a member of the IEEE and the Institute of Electronics, Information and Communication Engineers in Japan.

The Artech House Radar Library

David K. Barton, *Series Editor*

Modern Radar System Analysis by David K. Barton

Introduction to Electronic Warfare by D. Curtis Schleher

High Resolution Radar by Donald R. Wehner

Electronic Intelligence: The Analysis of Radar Signals by Richard G. Wiley

Electronic Intelligence: The Interception of Radar Signals by Richard G. Wiley

Pulse Train Analysis Using Personal Computers by Richard G. Wiley and Michael B. Szymanski

RGCALC: Radar Range Detection Software and User's Manual by John E. Fielding and Gary D. Reynolds

Over-The-Horizon Radar by A.A. Kolosov, *et al.*

Principles and Applications of Millimeter-Wave Radar, Charles E. Brown and Nicholas C. Currie, eds.

Multiple-Target Tracking with Radar Applications by Samuel S. Blackman

Solid-State Radar Transmitters by Edward D. Ostroff, *et al.*

Logarithmic Amplification by Richard Smith Hughes

Radar Propagation at Low Altitudes by M.L. Meeks

Radar Cross Section by Eugene F. Knott, *et al.*

Radar Anti-Jamming Techniques by M.V. Maksimov, *et al.*

Radar System Design and Analysis by S.A. Hovanessian

Aspects of Radar Signal Processing by Bernard Lewis, Frank Kretschmer, and Wesley Shelton

Monopulse Principles and Techniques by Samuel M. Sherman

Monopulse Radar by A.I. Leonov and K.I. Fomichev

Receiving Systems Design by Stephen J. Erst

High Resolution Radar Imaging by Dean L. Mensa

Radar Detection by J.V. DiFranco and W.L. Rubin

Handbook of Radar Measurement by David K. Barton and Harold R. Ward

Statistical Theory of Extended Radar Targets by R.V. Ostrovityanov and F.A. Basalov

Radar Technology, Eli Brookner, ed.

The Scattering of Electromagnetic Waves from Rough Surfaces by Petr Beckmann and Andre Spizzichino

Radar Range-Performance Analysis by Lamont V. Blake

Interference Suppression Techniques for Microwave Antennas and Transmitters by Ernest R. Freeman

Signal Theory and Random Processes by Harry Urkowitz

Techniques of Radar Reflectivity Measurement by Nicolas C. Currie

SIGCLUT: Surface and Volumetric Clutter-to-Noise, Jammer and Target Signal-to-Noise Radar Calculation Software and User's Manual by William A. Skillman

Radar Reflectivity of Land and Sea by Maurice W. Long

Aspects of Modern Radar, by Eli Brookner, *et al.*

Analog Automatic Control Loops in Radar and EW by Richard S. Hughes

Introduction to Sensor Systems by S.A. Hovanessian

VCCALC: Vertical Coverage Plotting Software and User's Manual by John E. Fielding and Gary D. Reynolds

Electronic Homing Systems by M.V. Maksimov and G.I. Gorgonov

Principles of Modern Radar Systems by Michel H. Carpentier

Secondary Surveillance Radar by Michael C. Stevens

Detectability of Spread-Spectrum Signals by Robin A. Dillard and George M. Dillard

Radar and the Atmosphere by Alfred J. Bogush, Jr.

Space - Based Radar Handbook, Leopold J. Cantafio, ed.

Airborne Pulsed Doppler Radar by Guy V. Morris

Multitarget-Multisensor Tracking: Advanced Applications, Yaakov Bar-Shalom, ed.

SACALC: Signal Analysis Software and User's Guide by William T. Hardy

Radar Reflectivity Measurement: Techniques and Applications, Nicholas C. Currie, ed.

Multifunction Array Radar Design by Dale R. Billetter

Active Radar Electronic Countermeasures by Edward J. Chrzanowski